Environment at the Margins

Environment at the Margins

Literary and Environmental Studies in Africa

Edited by
Byron Caminero-Santangelo and Garth Myers

OHIO UNIVERSITY PRESS
ATHENS

Ohio University Press, Athens, Ohio 45701
www.ohioswallow.com
© 2011 by Ohio University Press

To obtain permission to quote, reprint, or otherwise reproduce or distribute material from
Ohio University Press publications, please contact our rights and permissions department
at (740) 593-1154 or (740) 593-4536 (fax).

Printed in the United States of America
Ohio University Press books are printed on acid-free paper ⊗ ™

20 19 18 17 16 15 14 13 12 11   5 4 3 2 1

Library of Congress Cataloging-in-Publication Data

Caminero-Santangelo, Byron, 1961–
 Environment at the margins : literary and environmental studies in Africa /
edited by Byron Caminero-Santangelo and Garth A. Myers.
   p. cm.
 Includes bibliographical references and index.
 ISBN 978-0-8214-1978-6 (pb : alk. paper)
 1. African literature (English)—History and criticism. 2. Ecology in literature. 3. Africa—
In literature. 4. Ecology—Africa. 5. Ecocriticism—Africa. I. Myers, Garth Andrew. II. Title.
 PR9340.5.C36 2011
 820.9'96—dc23
                                                                    2011018432

Contents

Acknowledgments

THE EDITORS would like to acknowledge the support of the Kansas African Studies Center and its National Resource Center grant for making possible the 2008 colloquium from which this book emerged. In particular, we wish to thank the center's staff members Emmanuel Birdling and Craig Pearman along with student assistants Ashley Depenbusch and Kyle Shernuk. We would also like to acknowledge the generous contributions made by the University of Kansas Department of English and the Hall Center for the Humanities' Nature and Culture Seminar to help fund the colloquium.

We would like to thank the graduate students from Byron Caminero-Santangelo's "Postcolonialism and Ecocriticism" seminar and Garth A. Myers's "African Studies" seminar for assiduously attending all the sessions of the colloquium and for their questions and comments: Ali Brox, Erin Conley, Dustin Crowley, Anna Gonzales, Angela Kordahl Rapp, Andrew Kuhn, Ann Martinez, Paula Prisacaru, Stephanie Scurto, Samantha Simmons, and Shelley Stonebrook (Byron's students) and Dylan Bassett, Emmanuel Birdling, Steph Day, Ryan Gibb, Ryan Good, Ang Gray, Megan Holroyd, Hilary Hungerford, Anton Menning, Makame Muhajir, John Ringquist, Shimantini Shome, and Luke Struckman (Garth's students). Many colleagues from the University of Kansas also attended the colloquium and offered insights that greatly enhanced our discussions, and we particularly wish to thank Glenn Adams, Hannah Britton, Greg Cushman, Liz MacGonagle, and Don Worster in this regard. Omofolabo Ajayi-Soyinka, Chris Conte, and Rick Schroeder presented excellent papers in the colloquium but opted out of publishing their pieces in this book; we want to thank each of them for their stimulating contributions to the workshop.

We acknowledge the strong support of Gill Berchowitz and the Ohio University Press for this book from the idea stage onward. Finally, as always, we would like to thank our wives, Marta Caminero-Santangelo and Melanie Hepburn, and our children, Nicola, Gabriel, Phebe, and Atlee, for their love and patience through the many years from our first conversations about a possible conference along these lines through the hours of copyediting and proofreading.

Introduction

Byron Caminero-Santangelo and Garth A. Myers

IN EARLY June 2009, Shell Oil Corporation agreed to pay more than fifteen million U.S. dollars to a group of ten Nigerian plaintiffs, most prominently the son of writer Ken Saro-Wiwa. The plaintiffs had accused Shell Oil of collaborating with the Nigerian military in the 1995 execution of Saro-Wiwa and eight other leaders of the Movement for the Survival of the Ogoni Peoples (MOSOP). MOSOP and Saro-Wiwa had tied their nonviolent advocacy for human rights in Ogoniland to highlighting the oil industry's devastating impacts on the ecosystem of their Niger Delta homeland. Although Shell Oil still dismissed the charges, its willingness to settle the case—critics argued that the company "bought their way out of a trial"—inevitably brought home the message that large multinational corporations can be brought to justice for violations of human rights and environmental devastation.[1]

Ken Saro-Wiwa was a well-known writer whose life's work focused on environmental justice. On the other side of Africa, Kenya's 2004 Nobel Peace Prize winner Wangari Maathai traveled somewhat the reverse path. She is an environmental activist and the founder of the Green Belt Movement, and her best-selling memoir, *Unbowed,* is highly regarded.[2] In the tragic murder of Saro-Wiwa and the triumphal ascent of Maathai, we see clearly the manner in which the literary and environmental have been prominently connected in Africa as well as the ways in which those two figures emphasize the link between environmental activism and social justice. African writers-as-environmentalists and African environmentalists-as-writers offer powerful alternative ways of understanding nature, conservation, and development, in contrast with dominant ideas of environment. It is at these complex intersections of the literary and the environmental in Africa where the initial impetus for this book is located.

This volume developed as the result of an interdisciplinary colloquium on literature and environment in Africa held in the spring of 2008 at the University of Kansas. Participants explored uses of literature and literary modes of analysis in the study of African environments by geographers, anthropologists, and historians as well as the application of theoretical frameworks and forms of knowledge drawn from geography, anthropology, and environmental history in the study of African and colonial literatures (primarily Anglophone). The two key questions that we focused on were how African literatures and modes of analysis drawn from literary studies might contribute to ways of reading the environment in the other disciplines and how African literary studies might productively draw from studies of African environments. These questions point to the need for dialogue across disciplines to develop better understandings of different discourses regarding African environments and people's relationships with them. In fact, a primary theme that cuts across the volume is dialogue, not just dialogue among disciplines but also dialogue among different visions of African environments and environmental change in Africa.

The need for such dialogue is pressing. More than a century of imperial and neoimperial attitudes and practices has resulted in intractable environmental problems as well as in the need for new kinds of environmental discourses. These attitudes and practices have been fostered in numerous kinds of texts, including literary texts. At the same time, African writers have often been keen spokespeople regarding the dangers of these texts and their environmental repercussions, as has been the case in so many other areas of political action. There is still much work to be done in terms of reading literature from and about Africa in relation to studies in other disciplines involving narratives of environment. Such work enables an understanding of how African literary texts intersect with larger social texts regarding African environments and their material implications. Those working in literary studies can become more familiar with work being done in other fields that traces environmental attitudes through a wide variety of texts and over long periods of time while taking into account what the different methodologies from other disciplines might bring to the study of literature. Those working in these other disciplines can learn from literary texts and explore how approaches informed by literature and literary theory might contribute to their work. For example, how might literary readings bring attention to the ways that language and formal features such as genre, plotting, and narration operate to construct and deconstruct meaning in different kinds of studies of African environments?

This latter question signals that our volume is part of the burgeoning work, typically termed *ecocriticism*, that brings together environmental and literary studies. An early and commonly cited source defines the term *ecocriticism* as "the study of the relationship between literature and the physical environment" when such study moves beyond treating the environment as background (setting) or symbol.[3] While ecocriticism remains closely associated with literary studies, the term *ecocriticism* is increasingly also used to denote work in other disciplines focused on issues of environmental representation (work often influenced by literary and critical theory). Ecocriticism has always had an interdisciplinary component, although the necessary relationship between ecocriticism and science (especially ecology) has been complicated, and is also closely associated with political advocacy and specifically with theorizing "about the place of literature in the struggle against environmental destruction."[4] Ecocritics seek to make their work relevant to efforts directed at understanding environmental degradation and finding less destructive ways of living with and within nature than those offered by the dominant modern ways of the world. As Lawrence Buell claims, "The success of all environmentalist efforts finally hinges not on 'some highly developed technology, or some arcane new science' but on 'a state of mind': on attitudes, feelings, images, narratives," all of which can be found in "acts of environmental imagination."[5]

Ecocriticism initially developed as a subfield in Anglo-American literary studies.[6] In increasing numbers in the past ten years, however, articles, edited collections, special issues of journals, and monographs have focused on the intersection of ecocriticism with postcolonial cultural studies.[7] Such work has been termed *postcolonial ecocriticism* and often emphasizes the similarities between the two fields of scholarship in terms of a sense of political commitment, interdisciplinarity, and the interrogation of capitalist development and progress. This type of work also focuses on the need for postcolonial studies to be more cognizant of ecocritical concerns: "Although ecocriticism overlaps with postcolonialism in assuming that deep explorations of place are vital strategies to recover autonomy, post-colonial criticism has given little attention to environmental factors" (DeLoughrey, Gosson, and Handley, 5). Even more emphatically, postcolonial ecocritics seek to push against the margins of American and British ecocriticism both to include more postcolonial texts in ecocriticism and to argue that postcolonial literature and theory can transform ecocriticism through increased attention to imperial contexts.

Almost all theorists working to develop postcolonial ecocriticism have noted tensions between postcolonialism and what Buell (*Future*, 8) calls "first-wave ecocriticism." Following what is chastised as the environmentalism of the affluent, first-wave ecocritics favor literary representations that focus on knowing, appreciating, identifying with, and protecting nature in a relatively pure state and/or on natural forms of belonging. First-wave ecocriticism has a tendency to erase histories of indigenous peoples, of colonial conquest, and of migrations that disrupted notions of wilderness and rooted dwelling. In his groundbreaking article "Environmentalism and Postcolonialism," Rob Nixon (236) notes the many ways that ecocriticism's "dominant paradigms of wilderness and Jeffersonian agrarianism" all too easily lead to "ecoparochialism" and "spatial amnesia" in which the histories of indigenous peoples and the shaping of places by transnational forces are suppressed. What we get is "an environmental vision that remains inside a spiritualized and naturalized national frame."

First-wave ecocriticism's historical erasure also extends to questions of language and representation, often castigating poststructuralism and historical materialism for their skepticism regarding claims of being able to represent nature in ways that escape political positionality. First-wave ecocritics embrace mimetic approaches to environmental representation with a focus on the ways that literary writing might break through culturally and politically inflected constructions of the environment to achieve a clear, unmediated reflection of the natural world and to give voice to nature. Such a position assumes that we can have knowledge and representation that moves outside the shaping effects of culture and history and that language can become a lens through which we see the world rather than a code that organizes and gives meaning to it. When combined with first-wave ecocritic's valorization of ecology, this position can lead to an uncritical approach to Western science and its claims of scientific objectivity. For the postcolonial critic, a theoretical stance that denies that all modes of knowledge production entail "institutionalized ways of seeing with *histories*" is extremely problematic.[8] For example, such a stance can unwittingly justify the violence done to indigenous peoples, cultures, forms of knowledge, and places through an imperialism working in the name of objective science.

Efforts to make ecocriticism more responsive to historical relationships of power, to colonial history and its effects, and to cultural difference have been central to postcolonial ecocriticism, which emphasizes both the inextricable intertwining of cultural, political, and natural history and "the role of mediation in representing the environment."[9] Susie O'Brien ("Back," 194) notes that postcolonial theory, because of its focus on undermining colonialism's drive for "an unmediated possession of the world," highlights "the contradictions that inhere not just between, but also within, all putatively representational discourses, thereby pointing up the dangers of heeding claims by *any* cultural structures (including postcolonialism and ecology) to reflect the world transparently." Anthony Vital ("Toward," 90) focuses on the need to balance the assumption that "language constructs our apprehension of the material world ('nature')" with the recognition that language itself is "always mediated by culture and society." In other words, we need to acknowledge not only that language shapes our perception and understanding of the environment rather than giving us a transparent view of the environment but also that language itself is the product of social processes. As a result, all representations of the material world are situated; they give viewpoints of the world that are historically, politically, and culturally positioned. Both O'Brien and Vital ("Toward," 90) are especially concerned that ecocritics recognize "the historicity of ecology as modern science," including both its roots in colonial history and its more contemporary universalizing and potentially colonizing impulses. The goal is not to erase ecology's counterhegemonic or even anticolonial potential but instead to note how ecology (as discourse) has been rendered ambivalent through its history.

In their recent book *Postcolonial Ecocriticism*, Graham Huggan and Helen Tiffin (15), drawing from Vital, argue that postcolonial ecocritical work needs to "explore 'how different cultural understandings of society and nature'—understandings necessarily inflected by ongoing experiences of colonialism, sexism, and racism—have been deployed in specific historical moments by writers in the making of their art." From such a general goal come more specific tasks such as exploring the transformation of genres "in different cultural contexts," tracing "how postcolonial writers from a variety of regions have adapted environmental discourses," and "demonstrating the knowledge of *non*-western (non-European) societies and cultures." In many ways, this kind of postcolonial ecocritical work can be linked with what Buell refers to as second-wave ecocriticism, which fo-cuses on the positionality of environmental representation and knowledge and as a result has expanded ecocriticism and embraced the sort of cross-cultural dialogue that we seek in *Environment at the Margins.* However, postcolonial ecocriticism brings attention to both global imperial contexts and parts of the world often elided even by second-wave ecocritics, whose expertise remains predominantly in American and British literature.

This volume can be considered part of postcolonial ecocriticism. While assuming that environmental representation is always shaped by social history, the contributors address how a wide range of texts deploy imperial environmental discourses and/or alternative narratives in varied historical and geographical contexts across the continent, in histories, and in discursive and narrative strategies, from Garth Myers's discussion of colonial environmental discourse in Eric Dutton's *The Basuto of Basuto-land* to Mara Goldman's analysis of Maasai oral environmental dialogue. Just as important, the contributors often explore the transformation of existing tropes, genres, and concepts (including ecocritical concepts) or the significance of suppressed environmental epistemologies for reimagin-ing development, environmental protection, sustainability, and relation-ships between humans and nonhuman nature toward the goal of forging a better future for Africa. This journey in the book cuts an arc from Byron Caminero-Santangelo's and Anthony Vital's postcolonial ecocritical dis-cussions of Nadine Gordimer and John Coetzee, respectively, through analyses of intersections between literary and policy devices in the works of Ngugi wa Thiong'o and Zakes Mda, Mia Couto, Ben Okri, and Wangari Maathai in chapters by Laura Wright, Amanda Hammar, Jonathan High-field, and Rob Nixon, respectively.

Yet the volume's Africa focus begs us to question the significance of geographical delimitation for postcolonial ecocriticism. Is there an African ecocriticism? If so, what is its relationship with the broader field? There has not been as much explicitly ecocritical work on Africa as there has been on, for example, Caribbean literature, and prior to this book there have been no published edited volumes or full-length studies.[10] In one of the first published discussions of African literature and ecocriticism, "Ecoing the Other(s): The Call of Global Green and Black African Responses," Wil-liam Slaymaker argued "that global ecocritical responses to what is hap-pening to the earth have had an almost imperceptible African echo" (138) and called for both African writers and critics to embrace what he saw as a global ecocritical movement: "African creative writers and literary or cultural scholars would benefit from the global environmental movement"

(140). While agreeing with the call for greater attention to environmental concerns by African critics, Byron Caminero-Santangelo questioned Slaymaker's overreliance on first-wave ecocriticism in judging if a piece of literature is properly environmental and claimed that ecocritical theory itself would need to be decentered if it was to be relevant in the context of African literature and criticism.

If only first-wave criteria are applied, there has certainly been little ecocritical literary writing from Africa. African writers have primarily addressed pressing political and social issues in colonial and postcolonial Africa. Concomitantly, in terms of environmental representation, these writers are concerned with lived environments, the social implications of environmental change, and the relationships between representations of nature and power. Certainly this is evident in even a cursory reflection on, say, the way that the environment figures into the works of Chinua Achebe, Ngugi wa Thiong'o, Nuruddin Farah, Ayi Kwei Armah, Ousmane Sembene, and many more African authors. These writers do not focus on nature in its pure state or on its preservation.

In contrast with such writers' approaches to nature, as Caminero-Santangelo points out, the practice of conservation in Africa has often been underpinned by ideas about a pristine nature that is threatened by indigenous environmental practice and in need of protection by those from the West with proper environmental sensibility. Erased by such a narrative are the extensive intertwined history of nature and culture in Africa and the creation of spaces of pure wilderness through the forced removal of those with long histories of inhabitation. Furthermore, the focus on nature and its preservation in the context of Africa can shift attention from social problems or make such problems secondary to conservation (especially of fauna). In this context, an ecocriticism based on principles from the environmentalism of the affluent will not find much traction in African literary studies. In an early article on ecocriticism and South African literary studies, Julia Martin made a similar point. She noted the tension between "a definition of environmental priorities that was perfectly in keeping with the . . . colonial project" (3) and the concerns of the "the majority of South Africans" who would see such priorities as "irrelevant, and even inimical, to the struggle for social and political justice" (1). In this context, she found it striking that there was "a rather uncritical focus on 'nature writing'" in British and American ecocriticism and pondered "if opening the canon to other voices" might "subvert the genre's foundations": "Is the nature of Third World environments likely to produce the texts of wilderness, forests, and the great outdoors with which we are familiar? I think of the difficulties of teaching Wordsworth to students from the townships" (4). Anthony Vital ("Toward," 88) likewise notes that "an African ecocriticism would differentiate itself from ecocriticism in the North, which has . . . either not felt compelled to engage with the consequences of European colonialism or found the available forms of postcolonial criticism to be inconsistent with ecocritical goals and strategies."

Caminero-Santangelo, Martin, and Vital are clearly making arguments similar to those offered by scholars theorizing postcolonial ecocriticism. However, they also point to the need to take into account the specificity of cultural, discursive, and material contexts in Africa; the ways that modernity has shaped Africa; and the kinds of local responses that have been engendered. Discussion of such contexts gestures toward possible ways that Africa might be thought of differently in terms of environment. In the Western imagination, Africa has been and still is framed as a singularity constituted by absence—of time, civilization, or humanity—and this image has served to legitimate the exploitation of places and peoples in Africa. Given the history of this representation as well as the continent's heterogeneity, it is tempting to dismiss any representation of Africa as a place as a fantasy, and a dangerous one at that. As Mbembe somewhat grandly proclaims, "There is no description of Africa that does not involve destructive and mendacious functions."[11]

However, this constructed geographical category of Africa has also taken on its own reality as a result of history. We cannot ignore, according to Ferguson, the ways that the imaginary category has been accepted as "real," and Africa has become what he refers to as a "place-in-the-world," where the "world" means an

> encompassing categorical system within which countries and geographical regions have their "places," with a "place" understood as both a location in space and a rank in a system of social categories (as in the expression "knowing your place"). . . . That "Africa" (however heterogeneous or incoherent such a category may be in the eyes of scholars) is such a "place"—that is, a socially meaningful, only too real, and forcefully imposed position in the contemporary world—is easily visible if we notice how fantasies of a categorical "Africa" and "real" political-economic processes on the continent are interrelated.[12]

Africa as a category may be a phantasm of colonial discourse, but imperialism past and present has also brought this phantasm to life. Mbembe

(237) himself is guilty of deploying this Africa and the "Africans" who inhabit it regularly in his work. Africa has become different from the rest of the globe, but this difference can only be understood properly in terms of a history of unequal global economic and political connections feeding off of and giving reality to an assigned geographical position.

The twinned notion of connection and difference as a means of characterizing what Ferguson calls "Africa talk" also has profound significance in terms of environmental degradation and protection. Global environmental problems—global warming, overfishing of oceans, disposal of toxic waste—have already deeply affected many Africans. Yet most Africans are not the primary sources of these problems, nor do many Africans generally benefit from the resource exploitation that engenders them. More localized problems too are often shaped by global factors that are difficult for many Africans to address, in particular the shaping of local political, cultural, and economic conditions by the legacies of colonialism and (neo)imperial capital.[13] Cycles of poverty resulting from these legacies have had substantial negative impacts on African environments, and in turn the resulting environmental conditions have been major factors in these vicious cycles.

Problems in environmental conservation also need to be thought of in regard to long-term historical global relationships and the ways they have structured local conditions. Conservation policy has often been determined by imperial representations of African environments and people, such as representations of the "true" Africa as a wilderness empty of people and of African environments threatened by local environmental practices.[14] Enabled by such representations as well as the notion of African nature's unique immensity and exoticism, colonial-style fortress conservation of megafauna (often with a band-aid of community conservation) is given particular prominence through the operation of wildlife nongovernmental organizations and the tourist industry working with African governments. What tends to be ignored is both (long) local histories of environmental interaction and global (structural) causes for degradation, often with devastating effects for local peoples and ineffective preservation efforts in the long term.

We would suggest that given the significance, environmentally and otherwise, of Africa as a place-in-the-world, it is important to pursue an African ecocriticism. We recognize that such a project will be part of postcolonial ecocriticism, even while it requires the latter to account for differences and contradictions resulting from variations in geographic scale. Postcolonial ecocriticism—like ecocriticism and postcolonialism more generally—needs to make connections across cultural and historical difference but in the pursuit of unity should also resist suppressing differences. In this sense, the best kind of postcolonial ecocriticism will avoid becoming associated with too narrow a set of theoretical commitments. Buell ("Future," 11) has claimed that "ecocriticism gathers itself around a commitment to environmentality from whatever critical vantage point." Something similar might be said of postcolonialism, using the term *antiimperialism* instead of the term *environmentality*. In fact, ecocriticism and postcolonialism run the risk of becoming obsolete precisely when they become associated too closely with a single "critical vantage point," such as environmentalism of the affluent (in the case of ecocriticism) or poststructuralism (in the case of postcolonialism). If postcolonial ecocriticism allows itself to become tied to overly specific theoretical positions—for example, a Derridean approach to speciesism or a strictly Marxist version of environmental justice—it will risk betraying its resistance to colonizing, universalizing forms of representation and its commitment to true dialogue among different narratives of nature and culture.

From another angle, what makes Africa as a place-in-the-world such a critical site for further developing postcolonial ecocriticism is the relevancy of questions of environmental governance. Governance is most commonly understood as a means of getting at the shifting power dynamics of decision making in an era when the roles of states are in flux and responsibilities over environmental management are ostensibly decentralized, privatized, or made participatory.[15] Decentralization, democratization, and privatization are typically perceived to go hand in glove in discursive tactics designed to project an image of local empowerment and enfranchisement; in actual practice, across the continent disempowerment and disenfranchisement are ironically the common result.[16] The resultant crises in governance spiral together with conflicts over natural resources, while programs for the decentralized, democratized, privatized, and participatory management of natural resources are often central flashpoints of governance failures.[17] For the dominant narrative voices on Africa's environmental problems, the crisis points are deemed too grand and important to belong to fictitious nation-states, so the governance over them goes to the global scale, with international agencies constructing notions of "local community participation" that fit their interests (Ferguson, *Global Shadows*, 42–43).[18] Rhetoric on local community empowerment is mismatched with policies that foster elite capture of that empowerment process and instrumentalizes the participation to meet the needs of international donors and Western conservationists.

We feel that there are several ways in which a postcolonial ecocritical literary imagination can open up these governance questions. The first way involves offering an appreciation of environmental governance policy documents as literature. The performativity of policy documents from agencies such as the World Wildlife Fund, the African Wildlife Foundation, the United Nations Environment Program, the United States Agency for International Development, or the Global Environmental Facility of the World Bank is rife with literary devices and discursive tactics.[19] The second way is represented in numerous chapters in this volume: the critical analysis of literary works that directly engage environmental governance. What sounds at first to be a deadening prospect actually turns out to be an immensely productive vein of inquiry that this book's contributors are among the first to mine.

With the chapters' wide range of theoretical and disciplinary perspectives, this volume represents postcolonial ecocriticism in its broadest and most inclusive sense. In fact, in many cases the contributors are not focused explicitly on issues and debates in ecocriticism, and none are concerned with defining either postcolonial or African ecocriticism. Instead, they primarily focus on the urgent need to explore the limitations of current means of understanding African environments and environmental problems and to develop alternatives. The first three chapters begin as critiques of colonialist constructions of nature and landscape. Garth Myers uses Timothy Mitchell's concept of enframing colonial discourse in combination with theoretical insights from disability studies and critical geography to analyze Eric Dutton's 1925 book *The Basuto of Basutoland*, showing how the text attempts to transform the landscape's "order without framework" into colonialism's "segmented plan," from the bedrock on up. Roderick Neumann uses Theodore Roosevelt's *African Game Trails* as a vehicle for exploring performative and textual approaches to landscape and national identity, arguing for continued critical inquiry into historical travel writings because of how they can set up collective geographical visions through the present day. Jane Carruthers traces the trajectory of writing about elephant hunting and management in South Africa from the near extermination of the species in the nineteenth century to the debates around culling that developed in the second half of the twentieth century.

The next three chapters articulate alternative visions of African environments from African orature and literature. Mara Goldman brings us an analysis of the oral literary culture of the Maasai, arguing that an understanding of "change and continuity in African environments has always involved storytelling," and yet African stories are often ignored or marginalized in constructions of the African environment. In analyzing the possibilities for counternarratives that can lead the way toward ecological justice in the intersection of literary and environmental imaginations, she argues for "moving beyond Western ideas of narrative itself." Amanda Hammar focuses on Mia Couto's Mozambican novel, *Sleepwalking Land,* emphasizing the literary landscape at its heart, a landscape that holds in its grains of sand both the horrors of war and the hopes for its end and that begins to blur the often naturalized divides between nature and culture, between inner and outer worlds, between the living and the dead, and even between science and art. Jonathan Highfield's analysis of Ben Okri's *The Famished Road* focuses on deforestation and its impacts on agriculture and foodways, particularly in terms of the concomitant loss of indigenous agroforestry and the loss of control over food production. In Okri's novel, colonial capitalism and processes of imperial globalization eat everything in their path, especially the forest, and severely impact people's abilities to feed themselves.

The lived environment of farms—so central to African worldviews—connects Highfield's chapter to David McDermott Hughes's chapter. The two chapters juxtapose an indigenous, spiritualized belonging with a manufactured settler notion of belonging. Hughes explores the effort by European settlers in Africa to grapple with their minority status and their sense of exile by establishing connections with landscapes rather than with social others. The difference between the settlers' places of origins—and the language developed to describe them—and African landscapes created a disjuncture that defied the settlers' efforts to belong.

Although the theme of literary engagements with environmental policy and governance appears in earlier chapters (agroforestry at the heart of Highfield's chapter, agricultural governance in Hughes's chapter, erosion policy in Myers's chapter, and hunting and savanna management policies in chapters by Neumann, Carruthers, and Goldman), it comes steadily closer toward the foreground in the four chapters with which the book closes. In his analysis of Coetzee's *Age of Iron*, Vital draws on positions taken on environmental justice and waste management and, in the process, adds essential social and political geographies to ecocriticism. Vital shows that the novel, as it wrestles with issues of ethics in a Cape Town of shack settlements and political violence, exposes how existing concerns with sustainability and waste management simply work within established systems, not acknowledging the complicity of a broadly considered social history in the damage to life. Caminero-Santangelo's chapter follows on smoothly from this point in exploring how two of Nadine Gordimer's novels, *The*

Conservationist and *Get a Life,* can be understood in terms of (and can also be used to critically interrogate) Buell's concept of the environmental unconscious. The chapter argues that in their activation of the environmental unconscious, the novels challenge existing distinctions between nature and politics, between the environment and ideology, including the kind of distinctions offered by Buell's own theorizing of environmental unconsciousness that privileges the environmental over the political as grounds for consciousness. The chapter's literary theoretical ambitions are tied directly to issues of environmental governance and policy through Gordimer's treatment of conservation and ecology and the way the novels work individually and together to draw attention to differing understandings of environmental problems at particular moments and over time.

Laura Wright focuses on Ngugi wa Thiong'o's *Petals of Blood* and Zakes Mda's *The Heart of Redness,* which deal with the struggles against capitalist modernity and its devastating environmental impact. Wright argues that both novels depict the danger of turning to notions of a return to precolonial tradition and to "prelapsarian—and imaginary—African Edens." Whereas Ngugi turns to a fairly dogmatic form of Marxism as a means of contesting the socioenvironmental destruction that comes with a corrupt development project, Mda overtly questions the effectiveness of forms of local empowerment such as cultural villages and ecotourism, which must still traffic in romantic notions of culture and nature. In the book's final chapter, Rob Nixon focuses on the challenge of raising public sentiment regarding slow-moving environmental calamities, calling for an exploration of rhetorical means of turning those disasters into dramatic stories that will engage the interest and memory of a public with a sensibility shaped by the spectacle. Nixon draws on the example of Wangari Maathai, who brought attention to the attritional threats posed by deforestation and desertification for Kenya and offered a material center for resistance to the neocolonial government and elites whose stripping of Kenya's trees for their own gain was symbolic of a wider process of undemocratic, corrupt theft of the country's future for individual gain.

Even with our scope narrowed to Africa, the volume's coverage is still inadequate. Authors who belong in such a volume—Chinua Achebe, Bessie Head, Nurrudin Farah, and Ayi Kwei Armah—are not included, and Francophone literature does not appear. Western Africa is underrepresented, and there is somewhat of a focus on white African writing. We are not alone in struggling with such problems. In a recent issue of *Safundi* that focused on ecocriticism and South African literature, all of the critical analysis is about white writers, and much of it focuses on issues of animals and animal rights. In her afterword, Jennifer Wenzel does not skirt the issue: "The elephant in the room . . . is of course the fact that ecocriticism can hardly be untouched by the structural imbalances that have plagued and troubled southern African literary and cultural studies as a whole."[20] Our book has a similar elephant problem, and we are left with a solution quite similar to that which Wenzel (130) offered: "As represented in the essays collected here, South African ecocriticism is explicitly postcolonial, cognizant of the critiques articulated by Rob Nixon and others, even if as yet unable to address or overcome them fully." In contrast with the issue of *Safundi,* our book is not entirely centered on South Africa or southern Africa, nor is it at all exclusively focused on white African writing. In fact, part of the work toward dialogue that this volume performs involves connecting different streams of African postcolonial ecocritical possibilities.

Besides that about environmental governance, another possible dialogue that runs among and within chapters is between colonial ways of understanding African environments and alternative narratives. A number of essays trace a colonial vision of the relationship between nature and culture in Africa through travel writing, hunting narratives, autobiographies, and historical documents (chapters by Myers, Neumann, and Carruthers). These colonial texts depict African environments as places of wilderness in which nature is identical with, absent of, or threatened by Africans themselves. This vision reinforces binaries of nature and culture, wilderness and civilization, and the human and nonhuman that encourage an instrumentalist approach to both places and people. Such colonial narratives and their postcolonial permutations have had damaging effects and have led to ineffective efforts at environmental protection or regeneration.

These effects point to the urgency of finding ways to reenvision African environments and human relationships with them that will result in new environmental practices; however, breaking from colonial narratives and especially their epistemological underpinnings has not been easy. As Laura Wright emphasizes, the effort to reach toward a pure and authentic cultural model that stands apart from the impact of the West can all too easily result in forms of cultural commodification that are tied to the very processes of environmentally unsustainable modernity they were intended to combat. The notion that a postcolonial vision entails a movement beyond binaries may be clichéd, but the essays in this volume suggest that in order to move forward, there must be a questioning of the margins created by hierarchical dichotomies that divide the human and animal, spaces of home and waste, city and country, nature and culture, and wilderness and history. These essays imply that if we are going to move beyond the legacies of colonial environmental discourses and practices in Africa, we are going to need to think much more about environment at the margins, where the margin is the interstitial space of those inherited binary divides.

Many of the chapters here also engage in the significant and challenging dialogue between the local and the global. Much work in postcolonial and African studies has emphasized the dangers of global (imperial) narratives that define, evaluate, and place the local while carefully separating the local from global historical processes (thus making it easier to define and place). This work has pointed to the need to redefine the relationship between the two, to think about how the local can disrupt such global visions by challenging their categories and the kinds of relationships they postulate as well as by suggesting how such global visions are themselves driven by local conditions and priorities. In terms of African environments, this entails questioning hermetic models of place and recognizing the ways by which places have been shaped and represented through global processes involving uneven political, economic, and cultural exchange. We must acknowledge that African places are never outside such processes. At the same time, we must think carefully about the ways in which African understandings of African places and of both the external and internal processes that have produced them are highly varied because of different geographical, cultural, and historical conditions.[21]

African studies has long resided at the margins of global literary studies, including in the field of ecocriticism. While Africa arguably may occupy something of a more central post in global environmental discourses, upon closer examination, particularly from the standpoint of literary criticism of those discourses, what we see is an imagined Africa occupying a place-in-the-world assigned to it in the age of imperialism: nature, absent of people, and Edenic, in need of salvation from its own inhabitants. This volume contests these marginalizing visions by bringing literary and environmental studies of Africa into robust interdisciplinary dialogue.

Notes

1. Ed Pilkington, "Shell Pays Out $15.5m over Saro-Wiwa Killing," Guardian.co.uk, June 9, 2009.

2. Wangari Maathai, *Unbowed: A Memoir* (New York: Anchor, 2006).

3. Cheryll Glotfelty and Harold Fromm, eds., *The Ecocriticism Reader: Landmarks in Literary Ecology* (Athens: University of Georgia Press, 1996), xviii.

4. Laurence Coupe, ed., *The Green Studies Reader* (New York: Routledge, 2000), 302.

5. Lawrence Buell, *Writing for an Endangered World* (Cambridge: Harvard University Press, 2001), 1–2.

6. For overviews of American and British ecocriticism, see Lawrence Buell, *The Future of Environmental Criticism* (Oxford, UK: Blackwell, 2005); Greg Garrard, *Ecocriticism* (New York: Routledge, 2004); and Ursula K. Heise, "The Hitchhiker's Guide to Ecocriticism," *PMLA* 121, no. 2 (2006): 503–16.

7. Cara Cilano and Elizabeth DeLoughrey, "Against Authenticity: Global Knowledges and Postcolonial Ecocriticism," *ISLE: Interdisciplinary Studies in Literature and Environment* 14, no. 1 (2007): 71–87; Elizabeth DeLoughrey, Renee Gosson, and George Handley, *Caribbean Literature and the Environment* (Charlottesville: University of Virginia Press, 2005); Graham Huggan, "'Greening' Postcolonialism: Ecocritical Perspectives," *Modern Fiction Studies* 50, no. 3 (2004): 701–33; Graham Huggan, and Helen Tiffin, *Postcolonial Ecocriticism: Literature, Animals, Environment* (London: Routledge, 2010); Graham Huggan and Helen Tiffin, "Green Postcolonialism," *Interventions: International Journal of Postcolonial Studies* 9, no. 11 (2007): 1–11; Rob Nixon, "Environmentalism and Postcolonialism," in *Postcolonial Studies and Beyond,* edited by Ania Loomba, Suvir Kaul, Matti Bunzl, Antoinette Burton, and Jed Esty, 233–51 (Durham, NC: Duke University Press, 2005); Susie O'Brien, "Articulating a World of Difference: Ecocriticism, Postcolonialism, and Globalization," *Canadian Literature* 170, no. 1 (2001): 140–58; Helen Tiffen, ed., *Five Emus to the King of Siam: Environment and Empire* (Amsterdam: Rodopi, 2007); Anthony Vital, "Toward an African Ecocriticism: Postcolonialism, Ecology and *Life & Times of Michael K,"* *Research in African Literatures* 39, no. 1 (2008): 87–106; Anthony Vital and Hans-Georg Erney, eds., *Postcolonial Studies and Ecocriticism,* Special issue of *Journal of Commonwealth and Postcolonial Studies* 13, no. 2, and 14, no. 1 (2006–7); and Laura Wright, *Wilderness into Civilized Shapes: Reading the Postcolonial Environment* (Athens: University of Georgia Press, 2010).

8. Susie O'Brien, "'Back to the World': Reading Ecocriticism in a Postcolonial Context," in *Five Emus to the King of Siam,* edited by Helen Tiffin (Amsterdam: Rodopi, 2007), 187.

9. Cilano and DeLoughrey, "Against Authenticity," 79.

10. For previous work, see Byron Caminero-Santangelo, "Different Shades of Green: Ecocriticism and African Literature," in *African Literature: An Anthology of Criticism and Theory,* edited by Tejumola Olaniyan and Ato Quayson (Oxford, UK: Blackwell, 2007), 698–707; Julia Martin, "New, with Added Ecology? Hippos, Forests, and Environmental Literacy," *Interdisciplinary Studies in Literature and Environment* 2, no. 1 (1994): 1–11; William Slaymaker, "Ecoing the Other(s): The Call of Global Green and Black African Responses," *PMLA* 116 (2001): 129–44; Anthony Vital, "Situating Ecology in Recent South African

Fiction: J. M. Coetzee's *The Lives of Animals* and Zakes Mda's *The Heart of Redness*," *Journal of Southern African Studies* 31, no. 2 (2005): 297–313; and Dan Wylie, "Elephants and the Ethics of Ecological Criticism: A Case Study in Recent South African Fiction," in *Re-Imagining Africa: New Critical Perspectives*, edited by Sue Kossew and Dianne Schwerdt (Huntington, NY: Nova Science Publishers, 2001).

11. Achille Mbembe, *On the Postcolony* (Berkeley: University of California Press, 2001), 241–42.

12. James Ferguson, *Global Shadows: Africa in the Neoliberal World Order* (Durham, NC: Duke University Press, 2006), 6–7.

13. See William Adams and David Hulme, "Conservation and Community: Changing Narratives, Policies & Practices in African Conservation," in *African Wildlife and Livelihoods: The Promise and Performance of Community Conservation*, edited by David Hulme and Marshall Murphree (Oxford, UK: James Currey, 2001), 9–23; Jonathan Adams and Thomas McShane, *The Myth of Wild Africa: Conservation without Illusion* (New York: Norton, 1992); Melissa Leach and Robin Mearns, eds., *The Lie of the Land: Challenging Received Wisdom on the African Environment* (Portsmouth, NH: Heinemann, 1996); and Roderick Neumann, *Imposing Wilderness: Struggles over Livelihood and Nature Preservation in Africa* (Berkeley: University of California Press, 1998).

14. David Anderson and Richard Grove, eds., *Conservation in Africa: People, Policies, and Practice* (Cambridge: Cambridge University Press, 1987); Dan Brockington, *Fortress Conservation: The Preservation of the Mkomazi Game Reserve, Tanzania* (Oxford, UK: James Currey, 2002); Jane Carruthers, "Nationhood and National Parks: Comparative Examples from the Post-imperial Experience," in *Ecology and Empire*, edited by Tom Griffiths and Libby Robin (Seattle: University of Washington Press, 1997), 125–38; Jane Carruthers, *The Kruger National Park: A Social and Political History* (Pietermaritzburg: University of KwaZulu-Natal Press, 1995); Jan Bender Shetler, *Imagining Serengeti: A History of Landscape Memory in Tanzania from Earliest Times to the Present* (Athens: Ohio University Press, 2007).

15. Garth Myers, *African Cities: Alternative Visions of Urban Theory and Practice* (Atlantic Highlands, NJ: Zed Books, 2011).

16. Patrick Chabal, *Africa: The Politics of Suffering and Smiling* (London: Zed Books, 2009); and Jesse Ribot and Phil Oyono, "The Politics of Decentralization," in *Towards a New Map of Africa*, edited by Ben Wisner, Camilla Toulmin, and Rutendo Chitiga (London: Earthscan, 2006), 205–28.

17. Bill Derman, Rie Odgaard, and Espen Sjaastad, eds., *Conflicts over Land and Water in Africa* (Oxford, UK: James Currey, 2007); Goran Hyden, "Governance and the Reconstruction of Political Order," in *State, Conflict and*

Democracy in Africa, edited by Richard Joseph (Boulder, CO: Lynne Rienner, 1999), 179–96.

18. See also Thomas Bassett and Donald Crummey, "Contested Images, Contested Realities: Environment and Society in Africa's Savannas," in *African Savannas: Global Narratives & Local Knowledge of Environmental Change,* edited by Thomas Bassett and Donald Crummey (Oxford, UK: James Currey, 2003), 1–30.

19. Timothy Mitchell, "The Object of Development: America's Egypt," in *Power of Development,* edited by Jonathan Crush (London: Routledge, 1995), 129–57, and Michael Cowen and Robert Shenton, "The Invention of Development," in Crush, *Power of Development,* 27–43.

20. Jennifer Wenzel, "Meat Country (Please Do Not Feed Baboons and Wild Animals)," *Safundi: The Journal of South African and American Studies* 11, nos. 1–2 (2010): 129–30.

21. Andrew Byerley, *Becoming Jinja: The Production of Space and Making of Place in an African Industrial Town* (Stockholm: Stockholm University Department of Human Geography, 2005).

Bibliography

Adams, Jonathan, and Thomas McShane. *The Myth of Wild Africa: Conservation without Illusion.* New York: Norton, 1992.

Adams, William, and David Hulme. "Conservation and Community: Changing Narratives, Policies & Practices in African Conservation." In *African Wildlife and Livelihoods: The Promise and Performance of Community Conservation,* edited by David Hulme and Marshall Murphree, 9–23. Oxford, UK: James Currey, 2001.

Anderson, David, and Richard Grove, eds. *Conservation in Africa: People, Policies, and Practice.* Cambridge: Cambridge University Press, 1987.

Bassett, Thomas, and Donald Crummey. "Contested Images, Contested Realities: Environment and Society in Africa's Savannas." In *African Savannas: Global Narratives & Local Knowledge of Environmental Change,* edited by Thomas Bassett and Donald Crummey, 1–30. Oxford, UK: James Currey. 2003.

Brockington, Dan. *Fortress Conservation: The Preservation of the Mkomazi Game Reserve, Tanzania.* Oxford, UK: James Currey, 2002.

Buell, Lawrence. *The Future of Environmental Criticism.* Oxford, UK: Blackwell, 2005.

———. *Writing for an Endangered World.* Cambridge: Harvard University Press, 2001.

Byerley, Andrew. *Becoming Jinja: The Production of Space and Making of Place in an African Industrial Town.* Stockholm: Stockholm University Department of Human Geography, 2005.

Caminero-Santangelo, Byron. "Different Shades of Green: Ecocriticism and African Literature." In *African Literature: An Anthology of Criticism and Theory,* edited by Tejumola Olaniyan and Ato Quayson, 698–707. Oxford, UK: Blackwell, 2007.

Carruthers, Jane. *The Kruger National Park: A Social and Political History.* Pietermaritzburg: University of KwaZulu-Natal Press, 1995.

———. "Nationhood and National Parks: Comparative Examples from the Post-imperial Experience." In *Ecology and Empire,* edited by Tom Griffiths and Libby Robin, 125–38. Seattle: University of Washington Press, 1997.

Chabal, Patrick. *Africa: The Politics of Suffering and Smiling.* London: Zed Books, 2009.

Cilano, Cara, and Elizabeth DeLoughrey. "Against Authenticity: Global Knowledges and Postcolonial Ecocriticism." *ISLE: Interdisciplinary Studies in Literature and Environment* 14, no. 1 (2007): 71–87.

Coupe, Laurence, ed. *The Green Studies Reader.* New York: Routledge, 2000.

Cowen, Michael, and Robert Shenton. "The Invention of Development." In *Power of Development,* edited by Jonathan Crush, 27–43. London: Routledge, 1995.

DeLoughrey, Elizabeth, Renee Gosson, and George Handley. *Caribbean Literature and the Environment.* Charlottesville: University of Virginia Press, 2005.

Derman, Bill, Rie Odgaard, and Espen Sjaastad, eds. *Conflicts over Land and Water in Africa.* Oxford, UK: James Currey, 2007.

Ferguson, James. *The Anti-Politics Machine: "Development," Depoliticization and Bureaucratic Power in Lesotho.* Cambridge: Cambridge University Press, 1990.

———. *Global Shadows: Africa in the Neoliberal World Order.* Durham, NC: Duke University Press, 2006.

Garrard, Greg. *Ecocriticism.* New York: Routledge, 2004.

Glotfelty, Cheryll, and Harold Fromm, eds. *The Ecocriticism Reader: Landmarks in Literary Ecology.* Athens: University of Georgia Press, 1996.

Heise, Ursala K. "The Hitchhiker's Guide to Ecocriticism." *PMLA* 121, no. 2 (2006): 503–16.

Huggan, Graham. "'Greening' Postcolonialism: Ecocritical Perspectives." *Modern Fiction Studies* 50, no. 3 (2004): 701–33.

Huggan, Graham, and Helen Tiffin. "Green Postcolonialism." *Interventions: International Journal of Postcolonial Studies* 9, no. 1 (2007): 1–11.

———. *Postcolonial Ecocriticism: Literature, Animals, Environment.* London: Routledge, 2010.

Hyden, Goran. "Governance and the Reconstitution of Political Order." In *State, Conflict and Democracy in Africa,* edited by Richard Joseph, 179–96. Boulder, CO: Lynne Rienner, 1999.

Leach, Melissa, and Robin Mearns, eds. *The Lie of the Land: Challenging Received Wisdom on the African Environment.* Portsmouth, NH: Heinemann, 1996.

Maathai, Wangari. *Unbowed: A Memoir.* New York: Anchor, 2006.

Martin, Julia. "Long Live the Fresh Air! Long Live! Environmental Culture in the New South Africa." In *Literature of Nature: An International Sourcebook,* edited by Patrick D. Murphy, 337–43. Chicago: Fitzroy Dearborn, 1998.

———. "New, with Added Ecology? Hippos, Forests, and Environmental Literacy." *Interdisciplinary Studies in Literature and Environment* 2, no. 1 (1994): 1–11.

Mbembe, Achille. *On the Postcolony.* Berkeley: University of California Press, 2001.

Mitchell, Timothy. "The Object of Development: America's Egypt." In *Power of Development,* edited by Jonathan Crush, 129–57. London: Routledge, 1995.

Myers, Garth. *African Cities: Alternative Visions of Urban Theory and Practice.* London: Zed Books, 2011.

Neumann, Roderick. *Imposing Wilderness: Struggles over Livelihood and Nature Preservation in Africa.* Berkeley: University of California Press, 1998.

Nixon, Rob. "Environmentalism and Postcolonialism." In *Postcolonial Studies and Beyond,* edited by Ania Loomba, Suvir Kaul, Matti Bunzl, Antoinette Burton, and Jed Esty, 233–51. Durham, NC: Duke University Press, 2005.

O'Brien, Susie. "Articulating a World of Difference: Ecocriticism, Postcolonialism, and Globalization." *Canadian Literature* 170, no. 1 (2001): 140–58.

———. "'Back to the World': Reading Ecocriticism in a Postcolonial Context." In *Five Emus to the King of Siam,* edited by Helen Tiffin, 177–99. Amsterdam: Rodopi, 2007.

Ribot, Jesse, and Phil Oyono. "The Politics of Decentralization." In *Towards a New Map of Africa,* edited by Ben Wisner, Camilla Toulmin, and Rutendo Chitiga, 205–28. London: Earthscan, 2006.

Shetler, Jan Bender. *Imagining Serengeti: A History of Landscape Memory in Tanzania from Earliest Times to the Present*. Athens: Ohio University Press, 2007.

Slaymaker, William. "Ecoing the Other(s): The Call of Global Green and Black African Responses." *PMLA* 116 (2001): 129–44.

Tiffen, Helen, ed. *Five Emus to the King of Siam: Environment and Empire*. Amsterdam: Rodopi, 2007.

Vital, Anthony. "Situating Ecology in Recent South African Fiction: J. M. Coetzee's *The Lives of Animals* and Zakes Mda's *The Heart of Redness*." *Journal of Southern African Studies* 31, no. 2 (2005): 297–313.

———. "Toward an African Ecocriticism: Postcolonialism, Ecology and *Life & Times of Michael K.*" *Research in African Literatures* 39, no. 1 (2008): 87–106.

Vital, Anthony, and Hans-Georg Erney, eds. *Postcolonial Studies and Ecocriticism*. Special issue of *Journal of Commonwealth and Postcolonial Studies* 13, no. 2, and 14, no. 1 (2006–7).

Wenzel, Jennifer. "Meat Country (Please Do Not Feed Baboons and Wild Animals)." *Safundi: The Journal of South African and American Studies* 11, nos. 1–2 (2010): 123–32.

Wright, Laura. *Wilderness into Civilized Shapes: Reading the Postcolonial Environment*. Athens: University of Georgia Press, 2010.

Wylie, Dan. "Elephants and the Ethics of Ecological Criticism: A Case Study in Recent South African Fiction." In *Re-Imagining Africa: New Critical Perspectives*, edited by Sue Kossew and Dianne Schwerdt. Huntington, NY: Nova Science Publishers, 2001.

Chapter 1

"A Beautiful Country Badly Disfigured"

Enframing and Reframing Eric Dutton's The Basuto of Basutoland

Garth A. Myers

THIS CHAPTER is an analysis of Eric Dutton's 1925 book *The Basuto of Basutoland*.[1] I use Timothy Mitchell's concept in *Colonising Egypt* of an enframing colonial discourse, in combination with other theoretical insights, to analyze the book.[2] Dutton, who later worked as an administrator in four British African colonies, orchestrated the construction of Lusaka as the capital of the future Zambia, and authored four other books, began his colonial career in today's Lesotho in 1918–19. He used the experience as the basis for writing this text, mostly while recovering from multiple surgeries that followed his severe injury during World War I. *The Basuto* is a curious and fairly thin human geography of the country. The book is at once a patronizing misreading of the history of the Basotho and a searching attempt not only to understand the people but also to come to grips with the landscape. The text attempts to transform that landscape from what was an "order without framework," in Mitchell's terms, into colonialism's segmented plan from the bedrock on up (55). Yet Dutton has more interest in Basotho senses of place and cultural practices than in the broader colonialist canvas with which the book begins; he struggles with what he thinks he ought to say and what he likes and longs for in the African world around him.

This struggle between repulsion and desire appears to have been life-long. Dutton's book suggests ways in which British colonialism in Africa contained within it both attraction and repulsion for colonialists. This work also suggests the influences that came from Dutton's denial of his postwar disability. Even as he exemplifies tropes and tactics of colonialism's environmental order, Dutton struggled against daunting personal and physical demons with a humor that betrays both the sentimentalist imperial geographer and the survivalist tactics of a disabled man.

Eric Dutton

During Dutton's thirty-four-year colonial career, he engaged and corresponded intimately with the intellectual vanguard of British imperialism in Africa. He served in secretarial posts in five different British colonies in eastern and southern Africa between 1918 and 1952, and he published four geographically oriented books on Africa.[3] Like the works of better-known British geographers in Africa, Dutton's writings present him as "both accomplice in, and critic of, the business of imperialism."[4] The representations of place, landscape, and environment in his writings manifest this ambivalence.

Dutton, the youngest of nine children in a middle-class parson's family, was born in Yorkshire in 1895. Like all four older brothers, he entered the army after studying at Hurstpierpoint and Oxford. His first and only battle experience came at age twenty-one in 1916 at Gallipoli, where he suffered severe injuries to his legs and spine. Although he was one of the few officers or enlisted men of the North Yorkshire regiment to survive, he never regained full use of his legs and never lived without severe pain.[5] After Dutton experienced a half dozen surgeries, a long convalescence at home, and a brief attempt at a clerkship in Basutoland, Robert Coryndon, then the governor of Uganda, hired Dutton as his private secretary. Dutton served Coryndon in Uganda from 1920 to 1922 and then moved with him to his new post in Kenya. Dutton served in Kenya (1922–30) and moved on to serve in Northern Rhodesia (1930–37), Bermuda (1938–41), and Zanzibar (1942–52) before his retirement. It was during the early part of his service in Kenya that he finished writing his first book, *The Basuto*, and published it.

In this chapter, I give a close reading of this odd little book. From this book onward, Dutton's spatial sense of colonialism suggests how frequently a "sense of landscape" and the power to produce it went hand in hand in masculinist ideological justifications of imperial rule (*Kenya Mountain*, xi).

Yet Dutton had spent much of his time in Lesotho recovering from his horrific war wounds, which still affected his service in Kenya throughout his work on the book. As a consequence, the physical challenges of his experience make for a profound subtext to the book. Thus, throughout the book there is a bit of ambivalence about colonial power in Lesotho, as in much of Africa in the interwar years (1919–39). The context of its writing and the complexities of the author produce something quite other than a straightforwardly hegemonic masculinist imperial conquest narrative. Read broadly (meaning on a general level), *The Basuto* might seem to be a piece of "male megalomania," and we could stop there.[6] However, read more closely (meaning within the historical-geographical context of its production and with appreciation for author positionality), the book also speaks to important nuances of analysis. This means producing what James Duncan and Derek Gregory term "a principled recovery of the complex subject positions of both men and women" who authored colonialist geographies such as this while recognizing the "physical means through which they engaged them" within a particular moment in time.[7]

In this case, the 1920s produced some grave uncertainties about the imperial endeavor at home and abroad for Britain. An ambivalent as well as "hybrid and syncretic . . . yet unequal exchange" took place in British colonial discourse between British authors and African peoples in writing about and articulating Africa's geography in these decades, when colonial administrations were in place and apparently in charge.[8] Moreover, Dutton was one of hundreds of male British war veterans in Africa suffering from war-related disabilities that both impaired their physical capacities and affected the ways in which they were seen and treated by other men. Dutton's "physical means through which" he "engaged" the landscape of Lesotho (to appropriate Duncan and Gregory's phrase) become bodily manifestations of the character of colonial geographies in British Africa between the wars.

Colonial Enframing and Postexploration Landscape Geographies

The "siting, surveying, mapping, naming, and ultimately possessing" of colonial territory by Europeans in Africa as elsewhere depended upon geographical science.[9] Literary studies have recognized that "imperialist structures of attitude and reference" depended on "the way in which structures of location and geographical reference appear in the cultural languages of literature."[10] Much as Roderick Neumann does in chapter 2 in this book, I seek to extend the interplay between geographical and literary analyses of landscape in colonial African contexts.

Once the possessing, pacifying, and (re)naming of the landscape was done in the era of exploration geography, the spatial tactics of British colonialism in Africa broadened and deepened.[11] After World War I closed the Scramble for Africa and up until at least the 1960s, British authors churned out books that sought a popular audience on African geographical topics. Many authors of such works were paid employees of colonial administrations and were writing about the people with whom and places in which they lived and worked. Some of this work followed in the wake of popular imperialist travel writing, such as Theodore Roosevelt's *African Game Trails* (discussed in Neumann, chap. 2). Many but by no means all of these authors of this form of colonial geography were men for whom the "colonial tour" was a kind of rite of passage or assertion of belonging distinct from the agenda that Roosevelt brought to his writing.[12]

The postexploration colonial geographies of the interwar years in particular became part of a reorientation of ways of seeing the African political and ecological landscape in a more controlled frame, on a less fantastical map, beginning with books such as Dutton's. Mitchell (45–62) identifies several conceptual themes for understanding the professional spatial strategies of British colonialism's "enframing" order that sought to develop that sense of landscape, and at least one of Mitchell's themes is widely applicable to the colonial landscapes that these interwar works explored. This involved altering African "orders without frameworks" into "segmented plans" (44). As colonial administrations asserted their power into wider corners of the colonies and established a coherent order, it became steadily apparent that race, class, and gender segmentation was intrinsic to the plan of that order. To Mitchell (44), this was an essential part of colonialism's effort to separate the "container" (the colonizing power) and the "contained" (the African community). In practice, these plans of a carefully segmented order almost never worked, and actually the containers often became confused by the interwar period.

Here I assess Dutton's *The Basuto* as part of the enframing order in the British colonies of the interwar years. I attempt to show how he sought in the book to shape orders without frameworks into segmented plans and to differentiate container from contained. This emerges in repeated assertions of differentiation and British superiority. The mundane enframing and segmenting order has close links to the megalomaniacal masculinist geographies of the exploration era.

The Basuto points toward the cracks in this frame that appeared quite prominently by the interwar years. One uncertainty centers on the performance of masculinity. Militaristic ideas of manliness went hand in fist with the Scramble for Africa and British colonial enterprises there until World War I.[13] J. Bristow and R. Dixon point to "thrustingly masculinist . . . ripping yarns"—British narratives of African adventure in the late nineteenth century—as guides to this male ideal.[14] Christian ideology imparted to early twentieth-century British imperial masculine ideals, in equal parts, the muscularity that Berg, Bristow, and Dixon suggest and gentlemanliness: "the ideal of Christian manliness imagined a 'gentleman' equally at home in the public as well as the private sphere. A 'manly sensibility'—integrating robust manliness with refinement and tempering moral authority with a solicitous regard for dependents—would guide his conduct."[15] There were even considerable efforts to develop this "muscular Christianity" in African men living under British rule.[16]

Yet even this ideal was in practice highly unstable and variable, and it became more so after World War I. Colonial rule in many areas, particularly in Africa, was "haunted by a sense of insecurity" by the time of the interwar period.[17] This insecurity extended to the "preferred forms of masculinity."[18] Richard Phillips's detailed reading of the ebb and flow of masculinities in British adventure stories shows that the manliness on display in these stories is "not deterministic or static" but instead is decidedly plural.[19] But Phillips (86–87) also notes that "the geography of adventure spills over into . . . 'real' gendered subjects and spaces inspiring merchants, investors, travelers, settlers, and others." Connecting the analysis of discourse to its material implications for and context in the colonial enterprise is vital. One practical set of implications involved connections that texts by colonial officials had to "the practical tasks of building in brick and mortar"—that is, building the colonial order—and then to the growing sense that this order was not doing very well (Dutton, *Hyena*, 119). Another set of implications concerned the highly varied physical capacity to perform the hegemonic idea of colonial masculinity. In *The Basuto*, Dutton deployed a masculinist colonial discourse of spatial control and enframing alternated with the ambivalent, hybrid, and interdependent character of his encounter with colonized people.

Lesotho during 1918–19

At the time of Dutton's hobbled arrival in 1918, Basutoland had already been under British control in one arrangement or another for some fifty years and was nearing the end of its first decade of being completely encircled by the independent white-ruled Union of South Africa. Since 1884, Basutoland had been a Crown colony with a resident commissioner answerable to the British high commissioner in southern Africa. In a rather short span of time, Basutoland had also become "nothing more than a labor reserve for its powerful white neighbor."[20] Geographically and economically, Basutoland conformed precisely to what Samir Amin meant by "Africa of the labor reserves": a territory whose sole purpose in regard to the colonial system was to provide laborers for white settlers, in this case in South Africa.[21] Coryndon, the man who would later become Dutton's chief mentor in colonial administration, had just completed a contentious and rather unfulfilling two years as resident commissioner prior to Dutton's arrival. Coryndon was indelibly linked to the institutionalization of varied forms of indirect rule in his service as governor or resident commissioner in Basutoland, Swaziland, Barotseland (northwestern Rhodesia, the former Lozi kingdom), Buganda, and Kenya. As Bill Freund points out, these first three were "migrant labour reserve zones, a condition that went well with the depredations of 'traditional' authority reinforcing its own controls while acting in the colonial economic interests."[22]

Coryndon, and the British more generally, struggled to gain authority over the indigenous leadership of Basutoland in political terms. Economically the territory became completely under the thumb of white mining capital via contracted labor migration. Yet the Sotho paramount chief and his association of chiefs, and for that matter the "Sotho intelligentsia" of the Basutoland Progressive Association or the radical Lekhotla la Bafo movement that often opposed the traditional chiefs, seldom accepted the overarching authority of the British administration and indeed attempted to circumvent the resident commissioners in most matters.[23] For most of the first half century (i.e., 1884–1934) of the High Commission era in Basutoland, the British administration was preoccupied with the manipulation of internal political structures in Lesotho at the expense of social and political development. As J. Bardill and J. Cobbe rightly point out, this should not be characterized as neglect because such a characterization would imply that Lesotho failed to receive benefits in social or economic development that accrued to other British labor reserves in Africa, when in reality development can be hard to find in British Africa of the 1910s or 1920s or even later on with the Colonial Development and Welfare Act schemes of the 1930s and 1940s.

By the 1930s, more than half of the male population of Lesotho would be employed in South African mines and industries. The proportion in 1918 would not have been as high. Still, it is safe to conclude that as a result of massive labor migration to South Africa combined with the eviction of Sotho farmers from lands claimed by whites in surrounding areas of

South Africa after 1913, the Basutoland that Dutton saw in 1918–19 was in the middle of a long decline into underdevelopment, whether measured in self-sufficiency in agriculture, in soil quality due to overcrowding in the areas that could be farmed, or in the overall standards of living for most households.

Dutton served as an assistant clerk in Quthing District in southern Basutoland apparently for a long enough period to feel comfortable in writing a book about its people but not long enough to register much of any other impact on the place. Instead, physically lame as he was, all he really had the capability to complete was a short tour of the country, much of the time in the company of his brother Frederick, Basutoland's director of education.

Dutton's *The Basuto of Basutoland*

Dutton (*The Basuto,* 20) might have been describing his feelings about the people of Lesotho when he wrote of the Maluti mountains that "these strange mountains confront one; though they look like confusion personified, even they, in their anarchy, yield obedience to some mightier power and form up, however grudgingly, into colossal and majestic lines from North to South." To put this in Mitchell's conceptual terms for colonial spatial discourse, all of Basutoland, in its culture and its physical form, was an order without the framework that came with segmented plans in colonial spatial projects.

Making that order without framework into a segmented plan required, discursively, that Dutton establish a gendered, racial hierarchy. He began, as it were, with the bedrock. Remarking on the "grandeur" of Lesotho's parent rock, Dutton (18) writes that "it is necessary for the reader to grasp these details, these physical peculiarities . . . for they have had a marked effect on the history of the race and afford an excellent illustration of the close connection between the two sciences of history and geography. In short, they made Basutoland the asylum of the weaker races." To Dutton (25–28) as the disinterested imperialist observer, the "weakest" of these "races" was the "Bushman." He admits his is a "somewhat dark picture of the Bushmen," but this is a remarkable understatement considering his animalized and racist description of their "ungainly bodies" and notoriety as "great liars, great mimics, vindictive and cunning," living in "hunting packs" that "appear to have lived together as a herd."

Masculinist discourse is evident from the outset in Dutton's ethnographic science for establishing his racist hierarchy, which is itself very similar to that of Roosevelt, for example (see Neumann, this volume), in East Africa. Dutton's (25–28) link of patriarchy to civilization is baldly stated in his explanation of why the Bushmen cannot be described as living in tribes: the "tribe belongs to the patriarchal—as opposed to the savage—stage of society." He continues the theme in his delineation of eight cultural features that make the Basotho superior to their Bushmen forebears: the institution of permanent marriage, the tracing of descent through the male, paternal authority, the presence of religion, the domestication of animals, the holding of property, the cultivation of fields, and the use of a code of customary law (51). Dutton eschews a leering, sexualized portrayal of women, instead preferring to maintain a kind of Edwardian propriety. However, equating male dominance with civilization is a familiar theme in the book.

In spite of the certainty with which he falls into the patterns that Kearns, Pratt, and others identify with imperialist discourse, in *The Basuto* Dutton displays some ambivalence about what kind of an order he is joining.[24] The clashes of cultures wrought by colonial rule in Basutoland puzzle Dutton, and his footnotes and frequent asides allow the reader into his confusion. At one point he footnotes that "It is strange that, though one is constantly obliged to use the word, 'native,' the man himself knows no such word. It does not occur to him to classify himself under the abstract title which we are teaching him most dangerously to use" (14). Dutton considers overgeneralizations dangerous in *The Basuto* because of his understanding of the principals of white rule in southern Africa. While the ideological pull of colonialism held out for antihistorical othering or differencing tactics, the practical realities of administration by the interwar years demanded something more than a one-size-fits-all box for the native or primitive to be juxtaposed with the white civilized British.

Dutton (52–53) consequently warns that "it has been the habit of many writers to treat patriarchal society as if it were a Ford car turned out by standard . . . whereas, in fact, each tribe has clearly marked characteristics which make it a type of itself." The main types of "Bantu races" in southern Africa, to Dutton, are the Bechuana type and the Zulu type. Although this distinction of course fits comfortably with the accepted linguistic divide (even today) of Sotho-Tswana languages from Nguni (Zulu/Xhosa/Swazi) Bantu languages, Dutton's rhetorical purposes are well beyond objective science here, and they seem to bear the influence of Robert Coryndon, Kenya's governor, Dutton's employer at the time of writing. Youe (1986) argues that Coryndon thought quite highly of the Sotho people intellectually, in comparison to Nguni speakers, and hence in his hierarchy of "races" felt more hopeful about the eventual enlightenment of the Sotho in comparison with the Swazi (for whom he had also served as resident commissioner). For Dutton (55), the Zulu (Nguni) type, although "altogether better specimens physically," come in for criticism for their allegedly violent character. The Bechuana type, to which Dutton's (56) Basotho belong, may be "imitative" and taken by the "power of mimicry," but they are seen to have a "quickness, vivacity and good humor" that make them enjoyable company. It was no longer any good to be lumpers; imperialists had to become splitters. Dutton, in his late twenties, was learning how to split. Such comparisons are part of the colonial "world-as exhibition" that Mitchell (167) describes, "a world divided absolutely in two" (or, more accurately, segmented into several distinct parts) wherein the colonialist needs the native, natural other as a foil to establish the appearance of civilized order in a segmented plan.

It is not only in ethnographic turns in this book where Dutton the objective scientist and Dutton the imperialist are inseparable. When the author turns to the landscape and Lesotho's physical features, the pull of colonial ideology is ever-present. Dutton (24) comments at length on the problems of erosion in the countryside, arguing that "Basutoland has no more pressing question than this, and it will grow in urgency as time passes." He does not, again unsurprisingly, locate the genesis of the massive erosion that he photographs in the 1913 evictions from South Africa that led to a doubling of Lesotho's population or in the labor shortages induced by subsequent migrant absenteeism. Instead, here is where the antihistoricism of imperialist discourses hits pay dirt, so to speak. As Noyes put it, "Colonial discourse must construct a boundless, featureless, homogenous space which may serve as the stage upon which colonial desire may produce its fantasies."[25] The Basotho are fixed in time and yet somehow outside of the sweep of time, and as "sadly cut up and disfigured by erosion" as their "primitive" agriculture appears to Dutton (24) to cause the land to be, he seems to want to keep them there in Anne McClintock's (24) "anachronistic space." It is in Lesotho's countryside, and not its new colonial towns, where Dutton finds the settlement structure that he believes is fitting for the Basuto race.

Where the town and the location are termed a "travesty of European civilization," when one travels away from this "civilization," Dutton finds "well-ordered villages and a society of very ordinary laws" (13). Dutton appears to have more interest, affinity, and insight into Basotho homesites, senses of place, and day-to-day cultural practices than in the broader canvas with which the book begins of racist stereotyping and the selective tradition inherent in rewriting Basotho history. This difference between the bristling dismissiveness about African culture and history at the broad level and a friendly, intimate and sympathetic approach to the everyday world of African people runs through Dutton's published works, his correspondences, and government memoranda written by him. Some of this can be linked to the sentimentalism of imperialist discourse, but other aspects are linked to the context of the realities of minority administration under conditions of general financial duress and isolation from metropolitan theories of the colonial order. Many other administrators' memoirs show this bifurcation (e.g., Dundas, *African Crossroads*). And in these later passages of *The Basuto*, we can see some of Dutton's genuine fascination with African settlements begin to emerge.

In these sections, Dutton puts skin and bones to a discourse of "improving" the Basotho by indicating that they are a society not without virtues. At the same time, he shows empathy toward the subjects of his narrative. He appreciates and respects the traditions of village square town meetings in Sotho and Tswana customary democracy, even as the colonial administration was busy co-opting and circumscribing this customary democracy in both of the High Commission territories of Basutoland and Bechuanaland (today's Botswana). "The Mosuto dislikes nothing more than being denied or even begrudged his say," Dutton (58) writes, and at the end of meeting sessions, major issues are "talked over so long and by so many that the final verdict is really the general opinion of all present." The beer-and-work parties of Sotho farming help people "to get through the most toilsome work of the year" (61). Basotho dance, however much "Christian opinion looked askance at it" (72), intrigued Dutton especially in its symbolism, such as the "mimicry of the British soldier" (88) in male dances. Storytelling songs full of humor and bite about the hut-tax collectors and government clerks (such as Dutton) are not simply "merry" but instead highlight wisdom or flaunt or uphold conventions; folktales and local superstitions impress Dutton (72–88) with their "logic" as well as their "imagination." He is thus not entirely without awareness or appreciation of African discursive responses to colonialism.

In contrast to Dutton's near glorification early in the book of universal patriarchy in the abstract as the only progressive way of social power dynamics, when discussing its everyday dynamics in Lesotho he is both more observant and more sensitive. Male absenteeism resulting from mining has, to Dutton (75), made "wives . . . self reliant." Women show great skill in many spheres, with their time "fully occupied," while men are "lounging and gossiping" (71). He declares that Basotho women are mere "chattel" but then goes on to perceive "the natural eloquence of woman and her marked disinclination to regard herself as a chattel" (75). The Basotho are seen to be worthy models in their "hospitality and kindness to strangers" (75). Dutton writes as though he has surprised himself with the realization that "as with us, there are conventions to be observed" (81) in Basotho social relations. The young Dutton seems to be struggling between what he thinks he ought to say—the patronizing, racist, masculinist stereotypes of empire—and what he genuinely likes, admires, and even longs for in the African world around him. It is a struggle between repulsion and desire that appears to have been lifelong.

Dutton is most adept when discussing African sense of space and place. He first came to know African settlement patterns and architectural styles in Lesotho and admires the now justly famous house styles and residential wall engraving murals of Lesotho. Dutton is impressed with Basotho techniques of building "generally substantial" houses, many in stone, and their use of interior plaster; "ingenious designs executed with some skill" can be found inside and outside of the homes (63). He is astute enough here to connect design features and the substantialness of homes with the degree to which "nearly every able-bodied young man at some time or other leaves his home to earn money at the gold fields," even if he fails to recognize the overwhelming social hardships wrought by this dependency for Lesotho or the irony of the mural tradition being one that belongs to Sotho women. Furthermore, he sees as sensible the Basotho emphasis on the outside areas of houses, since "they do most everything outside, even cooking" (63). The cleanliness of homes, inside and out, catches his eye. He seems to especially enjoy how the *lelapa,* the open space around the front of the home, functions as "a sort of drawing room" (63), even as he misses the deeply spiritual bonds of the Sotho with the front courtyard.

Bill Freund (*The Making,* 147) reminds us that "colonialism was a phenomenon of overwhelming complexity that contained very attractive as well as repellent features for Africans." Dutton's book on Lesotho shows that this complex phenomenon contained within it both attraction and repulsion for the colonialists. His attraction to the quotidian details of African settlements and what he saw as Africans' "simple philosophy of life" (14) clashed with the overarching imperialist discourse. In his limited time in Lesotho, Dutton developed an understanding of the value placed on sharing and interdependence in village life, but he ultimately uses this not for a simple idealizing of rural Basotho life. He closes his description of village form with a Sesotho proverb, no doubt taught to him by his older brother, the education director: "*motse o motle kantle,*" meaning "the village looks pretty from the outside" (83). The proverb speaks volumes about Britain's colonial rule over Lesotho and the colonialist's vision of that rule. Dutton came to realize that there were "conventions to be observed" (81) for the Basotho, but he never realized that it might have been worthwhile to observe them himself. The colonial spatial project may have looked pretty and righteous from the outside, but the turbulence inside its villages and towns built steadily with time, often because of the clash of conventions right in front of Dutton's nose that he nevertheless failed to get inside because they were, like Basotho houses to him, "murky" (64).

The Basuto appears to have been partly a project encouraged upon Dutton by his mother in his many periods of convalescence. His mother and an uncle in fact did much of the referencing and created the book's index. His brother Frederick took most of the book's exceptionally stunning photographs and appears to have been the source for much of its detail about Lesotho. A. S. Owen, Dutton's mentor during his studies at Keble College, did the proofreading. The photographs are beautiful, the prose at times is unbearable, and the style is obfuscating, outrageous, and arrogant. In the end, the book can be read charitably as a volume of therapy, where much of the young Dutton's searching mind found refuge in a struggle to understand what was clearly an alien place and a difficult, even heartbreaking, experience for him.

Buried in the back of Robert Coryndon's papers in Rhodes House, which Dutton compiled and deposited for his mentor, I found a doctor's certificate filed on Dutton's behalf in 1919 just prior to his release from clerical duties in Quthing District. Because of Dutton's extensive rifle and shell wounds, the doctor pronounced him "unable to continue his work owing to acute neuralgic pains in both legs and in the spine, and septicemia from compound fractures in both legs." This piece of paper might have been misplaced by an archivist or reader, but I tend to doubt it; the paper was in the same place in 1996 where it had been in 1992. I admit that I never asked to be sure, but if my hunch is right, I think that Dutton put it there himself. What a funny place for Dutton to have put this little piece of paper, since his own memoir and some other papers and speeches of his are housed elsewhere at Rhodes House (he also compiled the papers and speeches of Coryndon's successor as Kenya's governor, Edward Grigg). Dutton was too careful and cunning as a man of discursive tactics to have simply mislaid this letter in Coryndon's papers. And Dutton didn't even know Coryndon in 1919. So, was he hiding his disability, or was he hiding his failures in Basotholand? Certainly his memoir, *The Night of the Hyena*, makes no note of his time in the future Lesotho, and he mentions himself nowhere in *The*

Basuto. But his memoir also makes no mention of Zanzibar, even though every iota of evidence in his life of letters and memoranda points to his eleven years there as the pinnacle of his career. The conclusion I reach is that his denial of the disability (a common theme in psychological analyses of the amputated, paralyzed, and severely disabled veterans of World War I) might have caused the deliberate misplacement of the doctor's note, but when mixed with the inveterate secretary's obsession with the preservation of information, Dutton knew that it ought to be archived somewhere.

The Basuto also didn't sell very well, and, in contrast with Theodore Roosevelt's *African Game Trails,* for instance, has had little of an afterlife in postcolonial times (see Neumann, this volume). Jonathan Cape published *The Basuto* in 1925, but Dutton waited in vain to make any money from it. He had a wound pension and a war bonus, which in 1922 combined for about 95 pounds a year, and later a personal allowance from Governor Coryndon in Uganda and Kenya of 150 pounds a year, not pauper's wages but hardly a sizable salary even then. The clear reality that Dutton did not belong to the elites and aristocrats who ruled the empire and its Colonial Office prefigured much of his ambivalent experience with that hierarchy in Uganda and Kenya and later in his career. In *Lillibullero* (13) he wrote of the "two races of mankind" as "the great races of the borrowers and the lenders," and he placed himself high among the borrowers and "non-payers." Financially strapped as he was until his appointment in Northern Rhodesia assured him an actual salary, he wrote continually to Jonathan Cape about the slow sales. In February 1924 Cape finally replied with a message that almost any academic book author can instantly recognize: "Please do not continue to write as if we had a grudge against your book and were not trying to sell it. Your book is one which certain special people will want and they are sure to learn of it and will take steps to obtain it."[26] His brother Frederick offered good cheer in letters from Basutoland to him in London and then Uganda, saying he was lucky to leave the "vegetative" state of things in Lesotho.[27] A. S. Owen wrote him from Oxford to say that he was "astonished at your vigor" and to encourage him in more writing endeavors.[28] Dutton wanted to write a book on the Roman Empire, another on the history of war medals, and still another on travel writing in the ancient world. But he had a life to live and practical goals and eventually a fulfilling set of responsibilities that he attacked with the "gusto" that would have gone into the books if he'd had the time or physical strength.[29]

As a consequence of these characteristics, I find it hard to subscribe to a deconstructionist's death-of-the-author approach to Dutton's *The Basuto.* Even as he exemplifies many tropes and tactics of the colonial order, he struggled against some daunting personal and physical demons with a humor that was not simply the sentimentalist tactic of an imperial geographer but instead was the survivalist tactic of a disabled man and the entertainment tactic of a man who always wanted to be the center of attention from the side of the stage but was forever having to borrow his costume.

Lillibullero (107–8) contains a hilarious passage wherein a district officer in northern Kenya brings Dutton his old copies of the *Times Literary Supplement,* and Dutton gleefully reads a book review in which an author who wrote a savage review of *The Basuto* gets savagely reviewed himself. "Did ever a poor writer have a sweeter revenge? I chuckled wickedly over this castigation. And when my friends called me vindictive, I continued to do so, without shame." When pushed, though, he admits an "addiction to purple patches" in his writing: "The truth is that the world is now and again all the better for a splash of color." The purple sentimentality dampens all of his books but nearly sinks *The Basuto.*

The point of colonial rule to Dutton was to make order out of the chaos: the creation of segmented plans out of orders without frameworks. Cartography, more than any other science, served the cause of "an appearance of order" (Mitchell, *Colonising Egypt,* 163), since "putting regions on a map . . . exercised power in a pure subtle form—as the power to name, to describe, to classify."[30] Mapmaking and toponymy had an "overwhelming fascination" for Dutton: "The thought of settling for all time the course of a river, the position of a peak, or how a valley runs, has enticed us to undertake weary marches and squander whole days from out of our small store," he wrote (*Kenya Mountain,* 81). Dutton sought to mark Lesotho for the empire with its spatial code, its sense of landscape. To name, describe, and classify was to possess the place within the hegemony of a spatial "system."[31]

But many of the tactics of colonial discourse outlined by Kearns (1997) for manly turn-of-the-century adventure-geography writers—objectivity, sentimentalism, and antihistoricism—become unsettled in Dutton's hands. Later in his career, African conceptions of landscape and the natural world became infused with Dutton's own. In *Kenya Mountain* (44), he wrote: "I remembered with regret that I had laughed at a Basotho tradition that their mountain fortress, Thaba Bosiu, the Hill of Night, was so called because, when the Zulus under Moselekatse found it impregnable, they said it had grown in the night. God knows, I shall never laugh at anything again." This manner in which Dutton's colonialist sense of landscape was transformed by exchanges and encounters with African ideas followed him throughout his career and writings. When he returned to East Africa in

1942 after thirteen years of service in Northern Rhodesia and Bermuda, he wrote to his friend Frederick Lugard a succinct summation of this hybrid sense of landscape and belonging: "We have at last arrived back to East Africa. . . . All my early service was spent in these parts. I suppose I have come to regard them as my other home. Africa is a queer continent—once you get it in your blood, you can never get it out."[32]

Dutton says nothing about his legs in *The Basuto*, but what he does say elsewhere (in *Kenya Mountain*, 42) speaks volumes about what he must have experienced in Lesotho: "On our expedition [on Mount Kenya] in 1924 we walked as far as the hut, more than twenty miles from [base camp], in one march. It was my first attempt at anything more than two miles for nine years, and I found it a heavy strain. I was hampered by an iron splint on one leg (as still in 1926) and then encumbered by having to walk with two sticks: it was past eight at night when I stumbled into camp, beaten and distressed. . . . I must have fallen more than a score of times."

Dutton's other books and correspondences show a broader pattern of how both he and others viewed his disability. His instant dislike of his second boss in Kenya, Edward Grigg, came from the moment that Grigg "directed a doubtful look at my crutches. I had seen that look before" (*Hyena*, 106). While Dutton admitted "my infirmity resulted in there being yawning gaps in my knowledge" (*Hyena*, 85), he deeply resented having the unspoken power of "that look" determine what he could and could not do. Occasionally, his correspondents exclaimed that they were "astonished at [his] vigor" (Coryndon Papers, 1924), but most either looked away or, it seems, turned to him again with "that look."[33] Dutton's career stalled in the 1940s, just like his two attempts on the summit of Mount Kenya had stalled. In 1949 he wrote to Secretary of State Arthur Creech Jones plainly begging for a governorship: "now that there are rumoured to be several governorships falling vacant I hope that my claims will be favourably considered."[34] That they had not to date been considered, he told Jones, "I am sure you will agree . . . might well tempt a man to consider he had got a life-size grievance—particularly a man who has achievements to his credit and has been four times decorated." Jones himself did not reply, but his papers contain a letter from Gilbert McAllister on his behalf that alludes to "certain difficulties with" any possible Governorship "which he was not able to discuss with me," despite "Creech Jones' admiration for you [Dutton] and for your work" (Jones Papers, 1949). It is hard to read this exchange of letters and not think that the "certain difficulties," at least to some degree, had "that look" in them.

The varied capacity to perform the hegemonic version of masculinity, like Dutton's disability, was implicit in the workings of power, promotion, and privilege within British Africa. Dutton's (*Hyena*, 112) own ambivalence about the ideal—he admitted that he was "never a pre-eminently soldier-like figure at the best of times" even as he went off trying to climb mountains in leg irons—parallels an ambivalent attitude toward African men and ultimately toward colonialism. His sense of colonial landscapes carried both a drive to create order out of chaos and a recognition that it was like building "castles in Spain" to bother trying.

Conclusion

Eric Dutton, in published works and in many correspondences, provides us with opportunities to look inside the discursive practice of British colonialism in Africa. To place this claim in its historical-geographical context, *The Basuto* as a text worked to articulate the spatial order or enframing of British colonial rule in the interwar years. Dutton was seeking to reorient his readers' ways of seeing Kenya—the colony that he served while writing the book—as well as Basutoland as places in need of a segmented plan. Yet Dutton's objectivity, his sense of the righteousness and unquestionable prestige of colonial power, and his own masculinity appeared unsettled and contradictory. Colonialist enframing had cracks in its frames such as this in the interwar years. We see both his understanding of the sense of landscape and his drive for enframing in British colonialism in Africa at that time but also the ambivalence, uncertainty, and contradiction of colonial power and colonial masculinity. To extrapolate from this text to the British Empire in Africa writ large is dangerous. Nonetheless, I suggest that these two themes were common across the discourse of British colonialism in interwar Africa.

Notes

1. Eric Dutton, *The Basuto of Basutoland* (London: Jonathan Cape, 1925).

2. Timothy Mitchell, *Colonising Egypt* (Cambridge: Cambridge University Press, 1988).

3. In addition to Dutton's *The Basuto of Basutoland*, see his *Kenya Mountain* (London: Jonathan Cape, 1929), *The Planting of Trees and Shrubs* (Lusaka: Government Printers, 1935), and *Lillibullero, or The Golden Road* (Zanzibar: Privately published, 1944). Two copies of his unpublished memoirs *The Night of the Hyena* were deposited at the Rhodes House Library in Oxford, one by his son Charles and the other by his friend Elspeth Huxley.

4. E. Kroller, as quoted in Alison Blunt, *Travel, Gender, and Imperialism: Mary Kingsley and West Africa* (New York: Guilford, 1994), 52.

5. My discussion of Dutton's upbringing and youth here borrows from my two articles on Dutton: "Intellectual of Empire: Eric Dutton and Hegemony in British Africa," *Annals of the Association of American Geographers* 87, no. 1 (1998): 1–27, and "Colonial Geography and Masculinity in Eric Dutton's *Kenya Mountain*," *Gender, Place and Culture* 9, no. 1 (2002): 23–38. See also N. Steel and P. Hart, *Defeat at Gallipoli* (London: Macmillan, 1994).

6. Anne McClintock, *Imperial Leather: Race, Gender and Sexuality in the Colonial Contest* (New York: Routledge, 1995), 24.

7. James Duncan and Derek Gregory, *Writes of Passage: Reading Travel Writing* (London: Routledge, 1999), 3 and 5.

8. Clive Barnett, "Impure and Worldly Geography: The Africanist Discourse of the Royal Geographical Society, 1831–1873," *Transactions of the Institute of British Geographers* 23, no. 2 (1998): 240.

9. Nicholas Dirks, *Colonialism and Culture* (Ann Arbor: University of Michigan Press, 1992), 6; see also Robert Young, *Colonial Desire* (London: Routledge, 1995), and Thomas Bassett, "Cartography and Empire Building in Nineteenth-Century West Africa," *Geographical Review* 84, no. 3 (1994): 316–36.

10. Edward Said, "Secular Interpretation, the Geographical Element, and the Methodology of Imperialism," in *After Colonialism,* edited by Gyan Prakash (Princeton, NJ: Princeton University Press, 1995), 30.

11. Penelope Hetherington, *British Paternalism in Africa, 1920–1940* (London: Frank Cass, 1978), and Lucy Jarosz, "Constructing the Dark Continent: Metaphor as Geographic Representation of Africa," *Geografiska Annaler* 74B, no. 2 (1992): 105–15.

12. For examples, see R. Crofton, *Zanzibar Affairs, 1914–1933* (London: Francis Edwards, 1953); Charles Dundas, *African Crossroads* (London: Macmillan, 1955); and F. B. Pearce, *Zanzibar: The Island Metropolis of Eastern Africa* (London: Fisher Unwin, 1920).

13. Lawrence Berg, "Reading (Post)colonial History: Masculinity, 'Race,' and Rebellious Natives in the Waikato, New Zealand, 1863," *Historical Geography* 26 (1998): 117.

14. J. Bristow, *Empire Boys: Adventures in a Man's World* (London: Harper-Collins, 1991), 135, and R. Dixon, *Writing the Colonial Adventure: Race, Gender and Nation in Anglo-Australian Popular Fiction, 1875–1914* (Cambridge: Cambridge University Press, 1995), 1.

15. Graham Dawson, *Soldier Heroes: British Adventure, Empire and the Imagining of Masculinities* (London: Routledge, 1994), 65.

16. See the works of Eric Dutton's main mentor in life, Joseph Oldham: *Christianity and the Race Problem* (London: Student Christian Movement, 1924), and, with B. Gibson, *The Remaking of Man in Africa* (London: Oxford

University Press, 1931). See also John Bale and Joe Sang, *Kenyan Running: Movement Culture, Geography and Global Change* (London: Frank Cass, 1996).

17. Nicholas Thomas, *Colonialism's Culture: Anthropology, Government and Travel* (Princeton, NJ: Princeton University Press, 1994), 15.

18. Dawson, *Soldier Heroes*, 1.

19. Richard Phillips, *Mapping Men and Empire: A Geography of Adventure* (London: Routledge, 1997).

20. Christopher Youe, *Robert Thorne Coryndon: Proconsular Imperialism in Southern and Eastern Africa, 1897–1925* (Waterloo, Ontario: Wilfred Laurier University Press, 1986), 102.

21. Samir Amin, "Underdevelopment and Dependence in Black Africa: Origins and Contemporary Forms," *Journal of Modern African Studies* 10, no. 4 (1972): 503–24.

22. Bill Freund, *The Making of Contemporary Africa*, 2nd ed. (Bloomington: Indiana University Press, 1998), 121.

23. See J. Bardill and J. Cobbe, *Lesotho: Dilemmas of Dependence in Southern Africa* (Boulder, CO: Westview, 1985), and G. Van Wyk, *African Painted Houses: Basotho Dwellings of Southern Africa* (New York: Abrams, 1998).

24. Gerry Kearns, "The Imperial Subject: Geography and Travel in the Work of Mary Kingsley and Halford Mackinder," *Transactions of the Institute of British Geographers* 22, no. 4 (1997): 450–472, and Mary Louise Pratt, *Imperial Eyes* (New York: Routledge, 1992).

25. Jon Noyes, *Colonial Space: Spatiality in the Discourse of German Southwest Africa, 1884–1915* (Reading, UK: Harwood Academic, 1992), 182.

26. Jonathan Cape to Eric Dutton, February 15, 1924, Papers of Sir Robert Coryndon, Box 12/5, folio 62, Rhodes House Library, Oxford, UK (hereafter Coryndon Papers).

27. F. H. Dutton to E. A. Dutton, March 25, 1924, Coryndon Papers, Box 12/5, folio 63A.

28. A. S. Owen to E. A. Dutton, April 27, 1924, Coryndon Papers, Box 12/5, folio 63B.

29. Elspeth Huxley wrote in her mimeographed preface to Dutton's unpublished memoirs that "gusto was one of Eric's qualities."

30. Johannes Fabian, *Language and Colonial Power* (Cambridge: Cambridge University Press, 1986), 24.

31. Henri Lefebvre, *The Production of Space* (Oxford, UK: Blackwell, 1991), 11.

32. Eric Dutton, to Frederick Lugard, in Papers of Sir Frederick Lugard, Box 10, Oxford, Rhodes House Library, 1942.

33. A. S. Owen to Eric Dutton, 1924, Coryndon Papers, Box 12.

34. Eric Dutton, to Arthur Creech Jones, in Papers of Arthur Creech Jones, Box 7, Oxford, Rhodes House Library, 1949.

Bibliography

Amin, Samir. "Underdevelopment and Dependence in Black Africa—Origins and Contemporary Forms." *Journal of Modern African Studies* 10, no. 4 (1972): 503–24.

Bale, John, and Joe Sang. *Kenyan Running: Movement Culture, Geography and Global Change.* London: Frank Cass, 1996.

Bardill, J., and J. Cobbe. *Lesotho: Dilemmas of Dependence in Southern Africa.* Boulder, CO: Westview, 1985.

Barnett, Clive. "Impure and Worldly Geography: The Africanist Discourse of the Royal Geographical Society, 1831–1873." *Transactions of the Institute of British Geographers* 23, no. 2 (1998): 239–51.

Bassett, Thomas. "Cartography and Empire Building in Nineteenth-Century West Africa." *Geographical Review* 84, no. 3 (1994): 316–36.

Berg, Lawrence. "Reading (Post)colonial History: Masculinity, 'Race,' and Rebellious Natives in the Waikato, New Zealand, 1863." *Historical Geography* 26 (1998): 101–27.

Blunt, Alison. *Travel, Gender, and Imperialism: Mary Kingsley and West Africa.* New York: Guilford, 1994.

Bristow, J. *Empire Boys: Adventures in a Man's World.* London: Harper Collins, 1991.

Coryndon, Robert. Papers of Sir Robert Thorne Coryndon, Boxes 1, 3 and 12, collected by Eric Dutton, and including his correspondence. Oxford: Rhodes House Library, 1924.

Crofton, R. *Zanzibar Affairs, 1914–1933.* London: Francis Edwards, 1953.

Dawson, Graham. *Soldier Heroes: British Adventure, Empire and the Imagining of Masculinities.* London: Routledge, 1994.

Dirks, Nicholas. *Colonialism and Culture.* Ann Arbor: University of Michigan Press, 1992.

Dixon, R. *Writing the Colonial Adventure: Race, Gender and Nation in Anglo-Australian Popular Fiction, 1875–1914.* Cambridge: Cambridge University Press, 1995.

Duncan, James. "Sites of Representation: Place, Time, and the Discourse of the Other." In *Place/Culture/Representation,* edited by James Duncan and David Ley, 39–56. London: Routledge, 1993.

Duncan, James, and Derek Gregory. *Writes of Passage: Reading Travel Writing.* London: Routledge, 1999.

Dundas, Charles. *African Crossroads.* London: Macmillan, 1955.

Dutton, Eric. *The Basuto of Basutoland.* London: Jonathan Cape, 1925.

———. *Kenya Mountain.* London: Jonathan Cape, 1929.

———. *Lillibullero, or the Golden Road.* Zanzibar: Privately published, 1944.

———. *The Night of the Hyena,* Unpublished memoirs of Eric Aldhelm Torlogh Dutton, on microfilm. Oxford: Rhodes House Library, 1983.

———. *The Planting of Trees and Shrubs.* Lusaka: Government Printers, 1935.

Fabian, Johannes. *Language and Colonial Power.* Cambridge: Cambridge University Press, 1986.

Freund, Bill. *The Making of Contemporary Africa.* 2nd ed. Bloomington: Indiana University Press, 1998.

Hetherington, Penelope. *British Paternalism in Africa, 1920–1940.* London: Frank Cass, 1978.

Huxley, Elspeth. "Introduction." In *The Night of the Hyena,* unpublished memoirs of Eric Dutton. Oxford: Rhodes House Library, 1983.

Jarosz, Lucy. "Constructing the Dark Continent: Metaphor as Geographic Representation of Africa." *Geografiska Annaler* 74B, no. 2 (1992): 105–15.

Jones, Arthur Creech. Papers of Arthur Creech Jones, Box 7: Correspondence with Eric Dutton. Oxford: Rhodes House Library, 1949.

Kearns, Gerry. "The Imperial Subject: Geography and Travel in the Work of Mary Kingsley and Halford Mackinder." *Transactions of the Institute of British Geographers* 22, no. 4 (1997): 450–72.

Lefebvre, Henri. *The Production of Space.* Oxford, UK: Blackwell, 1991.

Lugard, Frederick. Papers of Sir Frederick Lugard, Boxes 9 and 10: Correspondence with Eric Dutton. Oxford: Rhodes House Library, 1928–42.

McClintock, Anne. *Imperial Leather: Race, Gender and Sexuality in the Colonial Contest.* New York: Routledge, 1995.

Mitchell, Timothy. *Colonising Egypt.* Cambridge: Cambridge University Press, 1988.

Myers, Garth. "Colonial Geography and Masculinity in Eric Dutton's *Kenya Mountain.*" *Gender, Place and Culture* 9, no. 1 (2002): 23–38.

———. "Intellectual of Empire: Eric Dutton and Hegemony in British Africa." *Annals of the Association of American Geographers* 87, no. 1 (1998): 1–27.

Noyes, Jon. *Colonial Space: Spatiality in the Discourse of German Southwest Africa, 1884–1915.* Reading, UK: Harwood Academic, 1992.

Oldham, Joseph. *Christianity and the Race Problem.* London: Student Christian Movement, 1924.

————. Papers of Joseph H. Oldham, Boxes 5, 6 and 7: Correspondence with Eric Dutton. Oxford: Rhodes House Library, 1925–38.

Oldham, Joseph, and B. Gibson B. *The Remaking of Man in Africa*. London: Oxford University Press, 1931.

Pearce, F. B. *Zanzibar, the Island Metropolis of Eastern Africa*. London: Fisher Unwin, 1920.

Phillips, Richard. *Mapping Men and Empire: A Geography of Adventure*. London: Routledge, 1997.

Pratt, Mary Louise. *Imperial Eyes*. New York: Routledge, 1992.

Said, Edward. "Secular Interpretation, the Geographical Element, and the Methodology of Imperialism." In *After Colonialism*, edited by Gyan Prakash, 21–39. Princeton, NJ: Princeton University Press, 1995.

Steel, N., and P. Hart. *Defeat at Gallipoli*. London: Macmillan, 1994.

Thomas, Nicholas. *Colonialism's Culture: Anthropology, Government and Travel*. Princeton, NJ: Princeton University Press, 1994.

Van Wyk, G. *African Painted Houses: Basotho Dwellings of Southern Africa*. New York: Abrams, 1998.

Youe, Christopher. *Robert Thorne Coryndon: Proconsular Imperialism in Southern and Eastern Africa, 1897–1925*. Waterloo, Ontario: Wilfred Laurier University Press, 1986.

Young, Robert. *Colonial Desire*. London: Routledge, 1995.

Chapter 2

"Through the Pleistocene"

Nature and Race in Theodore Roosevelt's African Game Trails

Roderick P. Neumann

> Again and again, in the continents new to peoples of European stock,
> we have seen the spectacle of a high civilization all at once thrust
> into and superimposed upon a wilderness of savage men and beasts.
> Nowhere, and at no time, has the contrast been more strange
> and more striking than in British East Africa.
>
> Theodore Roosevelt, *African Game Trails*

ON A summery March morning in 1909, former president Theodore Roosevelt stood on the deck of the German ocean liner *Hamburg* as preparations for its departure from the Hoboken, New Jersey, pier were completed. In the crush of thousands assembled to see him off, the gilt buttons had been cut from his Rough Rider overcoat, and his hat had been knocked from his head. He waved to the raucous crowd of well-wishers one last time before ducking into his specially fitted "imperial suite."[1] His son Kermit, three Smithsonian scientists, a double-barreled Holland & Holland rifle (dubbed the "Big Stick"), and two tons of baggage accompanied Roosevelt on board. They were bound for Africa.

It seemed that all of New York, if not the entire country, had spontaneously agreed to help Roosevelt stage his "plunge into the wilds" of Africa.[2] Experts and old Africa hands publicly weighed in on his chances of surviving the trip.[3] New York ministers prayed that "he may return to us again in safety" and for "the highest success in his great quest," the Italian Chamber of Commerce presented a bronze tablet, President William Taft offered a gold pocket rule, and Fort Wadsworth fired a twenty-one–gun salute as Roosevelt steamed out of New York Harbor.[4] The *New York Times* and the Associated Press kept Roosevelt's expedition in the public eye, publishing frequent dispatches from Africa up to the moment he "emerged from the jungle" nearly a year later.[5] The National Geographic Society invited him to speak upon returning, thereby drawing comparisons of his trip with Commander Robert Peary's North Pole expedition and Sir Ernest Shackleton's Antarctic exploration.[6]

The greatest publicity for the trip, however, originated from Roosevelt's own pen. He had contracted with the publishing house Charles Scribner's Sons to write a series of articles on his safari for *Scribner's Magazine* and to produce a popular book. Roosevelt, we should recall, was a larger-than-life celebrity at a time when the field was less crowded than today. He had just finished his second term as president and had arranged with the Smithsonian Institution to embark on an expedition to collect zoological specimens in Africa. The Age of Empire was at its pinnacle, the United States had emerged as an imperial power, and international mass tourism was in its infancy. In sum, he was wildly popular and a widely recognized authority writing on African people, nature, and landscape at a time when educated North Americans were hungry to learn about other lands and cultures and the relation of the United States to them. Roosevelt would instruct them.

In this chapter Roosevelt's writings help elucidate the significant role of imperial travel writing in establishing categories of people, places, and landscapes and arranging them in a global hierarchical relational order. Following in the path of critical studies of travel writing, this chapter explores how the ideas of race, nature, and national identity intersect in the construction of Roosevelt's imaginary geography of Africa. It is important to critically evaluate the categories and hierarchies created and reproduced in this literary genre because they continue to structure collective geographic imaginaries today. Roosevelt's writings, particularly *African Game Trails: An Account of the African Wanderings of an American Hunter-Naturalist,* demonstrate how some wildlife tourism in Africa constitutes a form of colonial reenactment, sometimes quite literally. Tourism is performative in that colonial reenactment works to maintain the boundaries, categories, and hierarchies established in imperial travel writing. My principal method is the discursive analysis of Roosevelt's published writings and private correspondence from archival collections and of present-day mass media and tourism advertising.[7]

Imperial Travel Writing

At the time Roosevelt signed his publishing contract and steamed off to Africa, calling oneself an imperialist was no more pejorative than calling oneself a progressive. Indeed, there could be significant overlap in the labels. The term *imperialism* is a late nineteenth-century neologism that explained global political and economic developments, including the retreat from international free-trade policies and the conquest and colonization of roughly one-quarter of the earth by, primarily, European states.[8] Many political figures of the period "were proud to call themselves imperialists," and some considered imperialism an important element of domestic social policy (Hobsbawm, 60 and 68). Imperialism was bound to domestic notions of patriotism and national identity. As Eric Hobsbawm (70) remarked, "the idea of superiority to, and domination over, a world of dark skins in remote places was genuinely popular." Travel was critical to generating knowledge of the territories of empire and producing and reinforcing racial hierarchies. Those whose circumstances did not allow such travel could experience territorial conquest and racial superiority vicariously through international exhibitions, natural history dioramas, and, most commonly, imperial travel writing.

In applying the modifier *imperial* to travel writing, I thus intend a specific meaning beyond simply the reports of Westerners traveling in the Orient. I am referring to a particular historical moment of modernity when an increasingly literate American citizenry was awakening to a new imperial world order and the country's leading role in the creation of that world order. The imperial awakening in American popular culture offers something of a counterpoint to the ambivalence and uncertainty in British colonial rule that Garth Myers traces in his chapter for this volume. Scholars typically mark the 1898 war with Spain as the beginning of America's imperial identity, but American business interests had been expanding overseas investments and marketing for many years prior.[9] Thus, popular curiosity about the peoples and places of the world had been building for some time. Publishers responded to the "frenzied interest" that the war sparked by mass marketing atlases and popular geographic descriptions, which worked to connect racial typologies to continental boundaries.[10] It is in this era that the National Geographic Society's membership numbers exploded as it took on its unique role, primarily through the publication of *National Geographic Magazine,* as a popularizer and booster of American imperialism. Hence, imperial travel writing, published in a variety of mass print outlets, had a particularly didactic form and purpose yet required an entertaining and at times titillating style that would appeal to a broad readership.

In my analysis of Roosevelt's contribution to this subgenre, I build on a foundation of work by historical and cultural geographers that explores how metropolitan "travel and exploration writing *produced*" the non-European world.[11] Edward Said's critical analysis of Orientalism has been central in much of this work, particularly his notion of "imagined geographies" of European empire and his Foucauldian exploration of power/knowledge in the discursive construction of European identity in opposition to the Oriental "Other."[12] I draw from this critical literature five analytical and methodological insights to guide my examination of Roosevelt's texts. First, much of the work emphasizes the importance of imperial travel writing in constructing the "domestic subject," especially racial, gender, and national identities (Pratt, 5). I am particularly interested here in the way that the imperial travel experience contributed to the construction of national identities at a historical moment in which race and nation "bore intertwined as well as competing systems of meaning."[13] Second, Duncan and Gregory stress attention to the "physicality of representation," suggesting that written texts can be productively analyzed alongside visual representations.[14] Additionally, attention to physicality reminds us that travel authors are "corporeal subjects moving through material landscapes" (Duncan and Gregory, 5) as Myers similarly emphasizes in his chapter of this volume. Third, Said's imagined geographies concept stresses the importance of the "pre-texts"—previous written accounts and visual representations that collectively discursively construct non-European people and territories—that authors carry with them on their travels, actually or figuratively.[15] Fourth, Gregory's idea of "scripting" highlights the performative aspects of travel and tourism. Gregory understands scripting as "a developing series of steps and signals" in travel and tourism that produces "a narrativized sequence of interactions through which roles are made and remade."[16] Finally, there is an emphasis throughout this critical literature on the continuities between the colonial past and the postcolonial present.[17] Classification schemes, hierarchies, and landscapes, once constructed, are difficult to dismantle, even more so when colonial reenactment, inherent in many wildlife tourism experiences, works to continually shore them up.

Imagining an African Adventure

Roosevelt's plans for his African adventure are rooted in his youthful fascination with the tales of Africa's great white hunters. Books by the likes of

Cornwallis Harris, *The Wild Sports of Southern Africa* (1839) and *Portraits of the Game and Wild Animals of Southern Africa* (1840), and Roualeyn Gordon-Cumming, *Five Years of a Hunter's Life in the Far Interior of South Africa* (1850), seem for many present-day readers mostly mind-numbing chronicles of wildlife slaughter. However, such books were inspirational for Roosevelt, especially Harris's, which, as Carruthers details elsewhere in this volume, established the persona of the African hunter-naturalist. The books provided Roosevelt with not only the stories and images to conjure Africa but also the models to emulate. He aspired to re-create these experiences himself and produce his own hunter-naturalist account. His African safari would itself be a reenactment.

No writer had more influence on Roosevelt's African imaginary than Frederick Courtney Selous, a Late Victorian adventurer, naturalist, soldier, big-game hunter, and author of numerous books including *A Hunter's Wanderings in Africa* (1881). Roosevelt (*Game Trails*, 3) considered Selous to be "the greatest of the world's big-game hunters." In 1897 while serving as assistant secretary of the navy, Roosevelt struck up a personal correspondence with Selous that continued through 1915.[18] After becoming president, Roosevelt entertained Selous at the White House, wrote the forward to one Selous's books, and proclaimed in a letter to Selous that "you stand absolutely alone as the arch-typical big-game naturalist."[19] As the end of his presidential term approached, Roosevelt wrote to Selous for help in planning his hunting trip to Africa. Selous obliged, selecting an outfitter in British East Africa, advising Roosevelt on detailed supply lists, and ultimately accompanying him for part of the trip.

His correspondence with Selous suggests Roosevelt's ambivalences, anxieties, and desires regarding his planned African adventure. He wants a head of every kind of game species and intends to ask the colonial governments for "certain privileges as to shooting rare animals or shooting in reserves" but feels compelled to remind Selous that "I am no game butcher."[20] Roosevelt writes to Selous that "I want to take a trip back to the Pleistocene" but also that "I should wish to travel so as to be comfortable."[21] Roosevelt, who viewed hunting as cultivating "vigorous manliness," tells Selous that "I feel a little effeminate in going so much better equipt than you on your East African trip of five years ago."[22] Roosevelt resisted hiring an outfitter and guide, worrying that he felt "like a Cook's tourist" rather than the Victorian adventurers he sought to emulate.[23] The correspondence also reveals how Roosevelt positioned Africa within his imperialist worldview. He imagined Africa in parallel to North America, comparing the Boer War in South Africa to the Mexican-American War for Texas whereby the latter proved that it was a "great advantage of civilization that the AngloAmerican should supplant the IndoSpaniard."[24] Africa was another stage on which the Anglo-Saxon race performed its expansive global destiny. Africa was another frontier, like the North American frontier, to be settled by a superior race for the good of all. Roosevelt would later repeatedly draw this parallel in recounting his own African encounter.

The Adventure Takes Shape

Roosevelt was widely known as a big-game hunter and had published extensively on his experiences. He fashioned himself also as a naturalist: a killer but also a keen observer of big-game habits and habitats in the mold of Cornwallis Harris. Roosevelt had established the hunter part of the hunter-naturalist persona by deed, but the naturalist part required some external recognition of his scientific authority. For this he enlisted the Smithsonian Institution and a variety of natural history museums around the country. As a former U.S. president and a former member of the Smithsonian Board of Regents, Roosevelt readily got the institution's attention. From the Smithsonian and other museums' perspectives, having in their dioramas a lion or an elephant shot by a celebrity of Roosevelt's stature was fabulous publicity. The Smithsonian negotiated a detailed list of specimens that they would like shot for their collections. The American Museum of Natural History, the University of California, and other institutions placed their own orders, including rare species already spiraling toward extinction. The formal agreement christened the trip the Smithsonian–Theodore Roosevelt Expedition. Three of the Smithsonian's naturalists—Edgar A. Means, Edmund Heller, and Alden Loring—joined the expedition, and Roosevelt's son Kermit became the official photographer.

There remained the matter of financing this arrangement, which would cost roughly $1.8 million in today's dollars. Given Roosevelt's popularity and record as an author, there was some excitement in the New York publishing world surrounding his African adventure. He entertained many offers but settled on the deal with Charles Scribner's Sons for $50,000 for twelve magazine articles plus royalties from the book. The Smithsonian organized a special fund and an appeal to donors that eventually collected $50,000 from a group of anonymous donors, many of whom, it was ultimately revealed, had served in Roosevelt's presidential administrations.[25] Finally, Andrew Carnegie donated $25,000. These combined sums adequately financed all aspects, ultimately requiring Roosevelt to contribute only half of the money from his *Scribner's Magazine* deal.[26]

Having obtained financing and a scientific imprimatur, Roosevelt turned his attention to the itinerary. Arriving by steamer in the British East Africa port of Mombasa, he would then take the Uganda Railway to Nairobi and the European settler estates of the surrounding highlands. Along with Selous, Roosevelt had also enlisted the help of another friend, Edward North Buxton, a British conservationist and experienced African hunter. One of Buxton's family members by marriage, Sir Alfred Pease, had a large estate in what came to be known as Kenya's "White Highlands," and Roosevelt arranged to spend several weeks there. Similarly, he would stay for a period at Juja Farm, a "luxurious bungalow" owned by another highlands settler, William N. McMillan.[27] From these comfortable bases, Roosevelt would venture out on hunting safaris in places such as Mount Kenya and Lake Naivasha. In such fashion he would make his way along the Uganda Railway to its terminus at Port Florence, taking about eight months. From there he would board a lake steamer to the northern shore of Lake Victoria, arrive at Entebe, and then travel by foot caravan to Lake Albert and hence downstream, eventually reaching the Nile steamer service at Gondokoro (figure 1). While Roosevelt was on safari, his expedition would require a caravan of two hundred African porters. Roosevelt himself would have six personal attendants: two "tent boys," two gun bearers, and two "horse boys" (*saises*). Upwards of sixty African game beaters were used on some of the hunts (Roosevelt, *Game Trails*, 88).

Roosevelt's itinerary suggests that he was not venturing into the Pleistocene landscape of his African imaginary. The Uganda Railway, completed in 1901 and running 585 miles from Mombasa to the Kavirondo Gulf of Lake Victoria, had been turning a profit since 1905. The railway's construction triggered a boom in African peasant cotton production, expanded the production of groundnuts and sesame, and facilitated European settlers' takeover of the highlands and the spread of coffee and sisal plantations.[28] The construction project brought rifles, pumps, construction materials, and steam engines flowing into the region to rapidly transform its ecology and economy. The railway itself had direct transformative effects, as forests were cut to produce ties and fuel. Passengers could observe from the train station an army of nine hundred African tree cutters.[29] Most significantly, the railway made East Africa available to tourists. Many tourists had preceded Roosevelt on his planned route, among them the wife of an American engineer who similarly traveled across the East African Protectorate to Gondokoro and down the Nile to Cairo. On the caravan route, where she observed "empty champagne and beer bottles" strewn about by previous tours, African porters "laid out her clothes," "carried portable baths," and

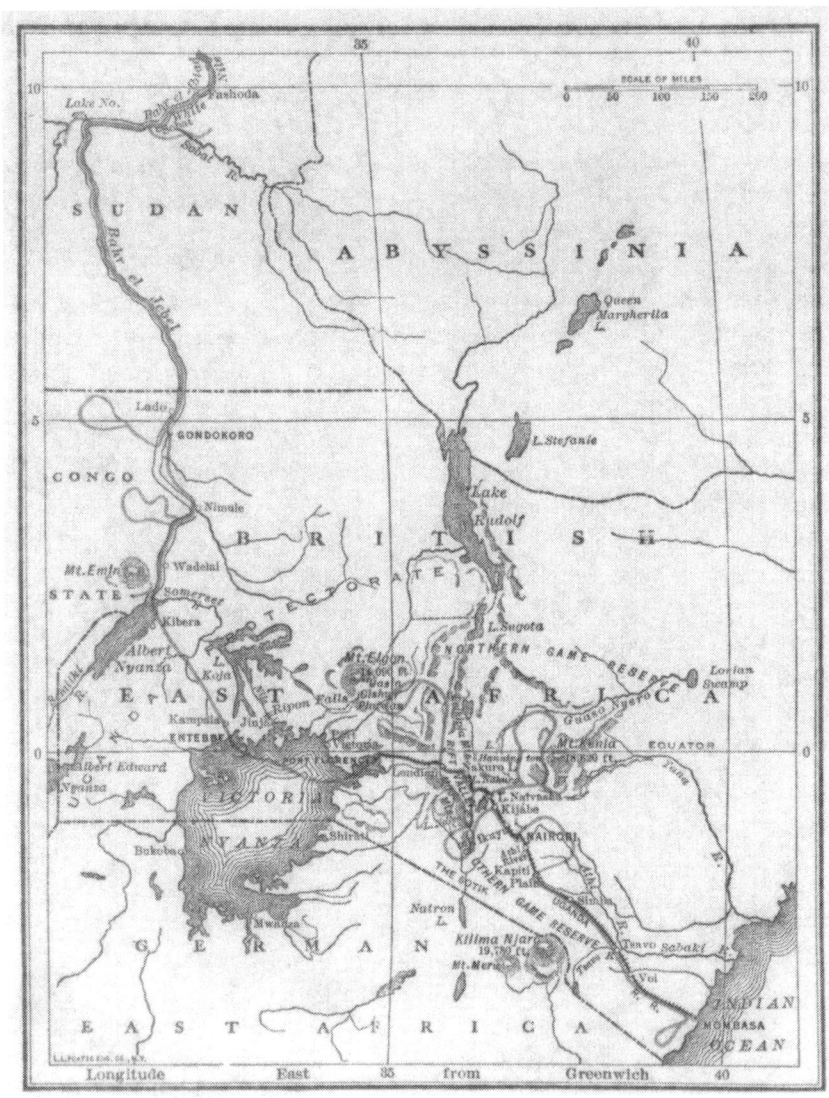

Figure 1. Photographic reproduction of the route map for the 1909–10 Smithsonian–Theodore Roosevelt Expedition. Courtesy of the Theodore Roosevelt Collection, Harvard College Library.

"prepared attractive dishes."[30] It is no wonder that Roosevelt was privately anxious that his African adventure plans were not matching his standards of masculinity.

Into the "Heart of Wild Africa"

Roosevelt (*Game Trails*, 1) entered the African "wilds" via the Uganda Railway or, as he titled his first chapter, on "a railroad through the Pleistocene." He took his place on the train not on the cushions of a luxury coach but instead on "a comfortable seat across the cow-catcher" (16), the view from which he represented as "literally like passing through a vast zoological garden" (19) (figure 2). His companions on the seat were Selous, the British East Africa governor Frederick Jackson, and Mearns. The visual spectacle of "Pleistocene" Africa was laid out before them, but the party was also visible for all along the railway to see. What more potent performance of imperial power, authority, and control over people and nature could there be than to strap the governor, the world's most renowned great white hunter, and the former U.S. president to the front of a steaming locomotive? Jackson and his metropolitan supervisors, Foreign Secretary Sir Edward Grey and Secretary of State for the Colonies Lord Robert Crewe-Milnes, granted Roosevelt virtual carte blanche for shooting anything with fur or feathers, including species nearing extinction, anywhere he liked, including game reserves. The party left the train for the ranch of Sir Edmond Pease, where it would spend the first two weeks.

For the next eight months, the expedition inched westward along the railway, staying in the houses of white highland settlers and in between making hunting forays of varying duration. Roosevelt and Kermit had two horses each at their disposal as well as all of the aforementioned porters and attendants. The hunting and collecting excursions were of two main types. There were daylong hunts on the various settler ranches where they stayed in which Roosevelt would venture out on horseback with few attendants. On Pease's land Roosevelt (63) encountered "scores of herds of the beautiful and wonderful wild creatures," and on McMillan's land Roosevelt shot a hippo within half a mile of the house. Sometimes these hunts were organized more as social events, with neighboring settlers and their wives all riding out on horseback to hunt with Roosevelt.[31] The other type of excursion entailed heading as much as 100 miles perpendicular from the railway for a period of two weeks or more. On these ventures the full complement of two hundred–plus porters and attendants were engaged not only to supply a luxury safari but also to support a scientific expedition. Four tons of

Figure 2. Theodore Roosevelt, British East Africa governor F. Jackson, F. C. Selous, and E. A. Mearns on the cowcatcher of a Uganda Railway locomotive. Courtesy of the Theodore Roosevelt Collection, Harvard College Library.

salt for preserving skins were loaded on to the train in Mombasa, and huge quantities from this supply were brought on long hunts. In a letter to his wife Edith, Roosevelt explained the "difficulty keeping our porters in food, for to collect huge animals for a museum means that a huge safari has to be taken along."[32] On these trips, Roosevelt had his own spacious tent with an extension for a hot bath. He found the tent "almost too comfortable," while he tolerated the hot bath as "almost a tropic necessity." Although he writes that his attendants "looked after my interest and comfort in every way" (*Game Trails*, 332), there is little else in his narrative that reveals what that might entail. Presumably he dined as comfortably as he slept on the extended hunting trips. In between visits with settlers and long safaris, the expedition would return to Nairobi to ship skins and specimens back to the museums.

Conditions for the expedition shifted notably following its departure for the Uganda Protectorate in late December 1909. Most of the hunting and collecting had been completed. The members of the expedition left the professional porters behind, bringing only their personal attendants. They traveled by steamer across Lake Victoria, stayed briefly in Kampala, and then headed out on a twelve-day march for Lake Albert. Every night on this leg they set up camp on the outskirts of the frequent African villages. Roosevelt and Kermit managed to hunt only two to three hours a day. They then sailed across Lake Albert and headed down the White Nile. Hunting was infrequent until they reached the next collecting ground, the Lado Enclave governed by the Belgian government. Here Roosevelt imagined himself to be in the "heart of wild Africa," writing to his sister Anna that he could see no settlements anywhere near them, the natives being "of a very primitive type, where there are any."[33] He noted that the Victoria Nile banks were "barren of human life," not because of the colonial government's forced evacuations but because here nature remains dominant, with "swarms of fly whose bite brings the torment which ends in death" (*Game Trails*, 454), keeping civilization at bay. Here also Roosevelt nudged the endangered northern white rhinoceros (*Ceratotherium simum*) closer to extinction, shooting six during the expedition's short stay in the Lado. Leaving their flotilla of small boats, Roosevelt and Kermit marched ten days downstream with their entire retinue to Gondokoro. They took their last extended hunt, crossing over the Nile to Redjaf with sixty Ugandan porters for a week to bag giant eland (*Taurotragus derbianus*). Finally, on February 28, 1910, they boarded the steamer for Khartoum, stopping whenever they sighted another specimen that they wished to shoot. Ultimately the Smithsonian received approximately five thousand mammal skins from the expedition, including one thousand large mammals. Including skins sent to other museums, the total rises to sixty-six hundred (an average of twenty animals per day) from Mombasa to Khartoum, excluding avian species.

In describing and photographically documenting the different phases of the expedition, Roosevelt presents a visual representation and spatial and corporeal ordering of race, nature, and national identity. "The [white] horsemen walked first, with the gun-bearers, *saises,* and usually a few very

Figure 3. The American flag was always carried near the front of Roosevelt's safari caravan. Courtesy of the Theodore Roosevelt Collection, Harvard College Library.

Figure 4. Theodore Roosevelt in camp, where the American flag always flew above his tent. Courtesy of the Theodore Roosevelt Collection, Harvard College Library.

energetic and powerful porters; then came the safari in single file; and then the lumbering white-topped wagons, ... each team led by a half-naked savage with frizzed hair and a spear" (180). Racial hierarchy was arranged spatially in the camps as well. "The tents were pitched in long lines, in the first of which stood my tent, flanked by those of the other white men and by the dining tent. In the next line were the cook tent, the provision tent, the store tent, the skinning tent, and the like; and then came the lines of small white tents for the porters" (98–99). In a nationalistic gesture in the heart of British Empire, a huge American flag was always posted in front of Roosevelt's tent and "was always carried at the head or near the head of the line of march" (95) (figures 3 and 4). The accompanying

Figure 5. Roosevelt and African attendants on a hunt near Redjaf, Lado Enclave, on the west bank of the upper Nile River. Courtesy of the Theodore Roosevelt Collection, Harvard College Library.

Figure 6. The expedition party standing over one of many African game kills. Courtesy of the Theodore Roosevelt Collection, Harvard College Library.

photographs confirm the domination of whites over African people and nature, positioning them in the foreground, mounted and elevated from the attendants and porters, or standing over the remains of the game they had killed (figures 5 and 6). Additionally, Scribner's hired for the book the well-known outdoor illustrator Philip R. Goodwin, who also illustrated Jack London's *Call of the Wild*. Drawn thousands of miles and many weeks distant from the events they depicted, Goodwin's illustrations were used to produce a heightened sense of adventure and danger. They also served to place Africans in a primitive state, subject to the ravages of nature (figure 7). Like many of the images produced on today's photographic safaris, Goodwin "cropped out" the white audience for whom the spectacle of a "native hunt" had been organized (406).[34]

Roosevelt's *African Game Trails* was didactic but in ways that uniquely reflected his positionality, the scripted rituals of the imperial hunt, and Victorian fields of knowledge. The book's title, an amalgam of his idols' earlier titles, establishes his identity nationally and as a sportsman and scientist, an American hunter-naturalist. Along with the internalized pre-texts of the

Figure 7. Philip R. Goodwin's rendering of Nandi warriors' lion hunt. Courtesy of the Theodore Roosevelt Collection, Harvard College Library.

great white hunters of the past, the party also carried a set of contemporary texts of the subgenre. "We had with us several recent books on East African big game; Chapman's 'On Safari,' . . . Powell Cotton's accounts . . . and Buxton's account of his two African trips. Edward North Buxton's books ought to be in the hands of every hunter everywhere . . . because they teach just the right way in which to look at the sport" (381). Roosevelt was thus explicit about the didactic value of the subgenre, but his own book, while in many ways following an established formula, went well beyond discoursing on the proper way to hunt. He was self-consciously a nationalist and an imperialist, one of the most important of the era, and his commentaries on game hunting and the order and qualities of nature are interspersed with, indeed inseparable from, his commentaries on racial hierarchies, national and racial traits, and racial destinies.[35] Much of the work of the classifying and ordering of races and nations had been accomplished in pre-texts, and Roosevelt (*Game Trails*, 9) was able to add his own authoritative voice to elaborating these schemata before he even set foot in the imagined "white man's country" of East Africa. Aboard the steamer the *Admiral* bound from Naples to Mombasa, he had already begun celebrating imperial expansion in Africa and the white "races" that were making it happen on the ground.

Frontiers and Pioneers of Empire

Many of Roosevelt's first recorded impressions, from his Indian Ocean passage to his first train ride, elaborate not on hunting or natural history but rather on the fulfillment of racial destiny. On ship he encounters the planters and government officials "going out to take command of black native levies in out-of-the-way regions where the English flag stands for all that makes life worth living" (4). He muses about the need to avoid conflicts such as the Boer War, preferring instead "friendly rivalry" between white nations destined for imperial rule. Following his understanding of environmental determinism, he argues that the white races share a common natural destiny to colonize the world's temperate zones. There is a natural order to be achieved through cooperation: "at least part of the high inland region of British East Africa can be made one kind of 'white man's country'; and to achieve this white men should work heartily together" (5). Progress and development in East Africa will "depend exclusively upon the masterful leadership of whites," and therefore "everything should be done to encourage such [white] settlement" (9).

Roosevelt's opinions about white settlement in East Africa were not founded on experience or extensive knowledge of the region but rather on a nationalistic and highly racialized narrative of European colonization of North America. This is made clear through his propensity to see the American western landscape and American pioneers wherever he looked in the East African highlands. At his first train stop in the Kitanga Hills (near Machakos), he found the place to be "a country of high promise for settlers of white race. In many ways it reminds one rather curiously of the great plains of the West, where they slope upward to the foot-hills of the Rockies." This impression of being in the American West quickly comes to dominate his reading of the landscape: "I might have been on the plains anywhere, from Texas to Montana; the hills were like our Western buttes." The parallels he drew between East Africa and the American West were the physical landscape as well as the unfolding of history written in the landscape. "There was much to remind one of conditions in Montana and Wyoming thirty years ago; the ranches planted down among the hills and on the plains still teeming with game, the spirit of daring adventure everywhere visible." And, most critically, he sees in the East African settlers planted among the hills the shared identity and imperial destiny of white races everywhere. He wrote of two "Africander" settlers in Kitanga: "From the first moment they and I became fast friends. . . . They reminded me, at every moment, of those Western ranchmen and homemakers with whom I have always felt a special sense of companionship." For Roosevelt, the narrative of world history is the narrative of the white pioneer, repeated "[a] gain and again, in the continents new to peoples of European stock" (*Game Trails*, 38, 39, 42, 79, and 1).

Everywhere Roosevelt looked, he saw only "the best kind of pioneer family" and "just the men for work in a new country." Early on he met two immigrants from South Africa who "represented the ideal type of settler for taking the lead in the spread of empire" and "typified in their lives and deeds the greatness of the English Empire." Blood equaled destiny, for these particular settlers "had never seen England." Other South Africans, Boers who were "as good a type as any one could wish for," fulfilled what Roosevelt considered "the three prime requisites for any race: they worked hard, they could fight hard at need, and they had plenty of children." He had a particular affinity for South African Boers, believing that they shared with him a common essential Dutch character and destiny for imperial expansion. This partly explains his ambivalence over taking sides in the Boer War and his preference for "friendly rivalry" and even intermarriage. Both the Dutch and the English were destined for pioneering, Roosevelt believed: "Both are so good that I earnestly hope they will become indissolubly welded into one people." To Roosevelt, "it was evident that the whole Uasin Gishu country would soon be occupied [by whites] and no better pioneers exist to-day than these South Africans, both Dutch and English." The only exception to Roosevelt's praise for whites in Africa was presented as a cautionary tale of racial transgression and miscegenation. Once "the most famous elephant hunter" in the region, Arthur Neuman, ended up "living in this far-off region exactly like a native, and all alone among the natives; living in some respects too much like a native." In the end, he "died by his own hand" (*Game Trails*, 178, 249, 26, 48, 405, and 364–65). The chasm between white civilization and African savagery was too vast to bridge.

Writing African Race and Nature

While Roosevelt had nothing but praise for white settlers, he had little but disdain for the Africans they were displacing. He placed almost all Africans at the bottom of his racial scale, best suited for subjugation by superior races. The "Wakamba . . . are in most ways primitive savages, with an imperfect and feeble social, and therefore military organization" (*Game Trails*, 43–44). He was in fact as preoccupied with race as he was with hunting, creating throughout the book elaborate if unsystematic racial categories and hierarchies. Roosevelt was not unusual in his "passion for classification" of both the natural and human worlds. "Indeed, it was the scientific and classificatory urges that helped to identify 'advanced' peoples" (MacKenzie, *Empire of Nature*, 39). Roosevelt's racial schema aligned with the standard cultural evolutionary model of the day. "Most of the tribes were of pure savages; but here and there were intrusive races of higher type; and in Uganda . . . lived a people which had advanced to the upper stages of barbarism" (*Game Trails*, 2). His rankings did take a peculiarly unique nationalistic character, however: "One of the Government farms was being run by an educated colored man from Jamaica; and we were shown much courtesy by a colored man from our own country who was practising as a doctor. No one could fail to be impressed with the immense advance these men represented as compared with the native negro." Additionally, according to Roosevelt's classification, our natives were more fierce than theirs. "Even the bravest of them, the warlike Masai, are in no way formidable as our Indians were formidable when they went on the warpath." Based on Roosevelt's racial hierarchy, the conquest of the American West was thus a greater achievement than Britain's conquest of East Africa. His ultimate conclusion for much of what he saw of African culture was that it "does not differ materially from what it was in Europe in the late Pleistocene" (*Game Trails*, 10, 42, and 2). East Africa was a land of Stone Age hunter-gatherers.

Roosevelt's racial classifications were related to the capacity to control nature. Africans were incapable of mastering nature and destructive of it, a contradiction that is never resolved. He sees mostly "wild pagans . . . unchanged in the slightest . . . during the countless ages when they alone were the heirs of the land—a land which they were utterly powerless in any way to improve." Africans were at nature's mercy because of their own cultural failings. In times of drought, "the foolish creatures die by the hundreds when they might readily be saved if they were willing to eat the herds. . . . Yet these savages are . . . wastefully destructive of the forests" (*Game Trails*, 18 and 44). In contrast, the racial superiority of European settlers is demonstrated by their mastery of nature. As detailed in the previous section, Roosevelt viewed Europeans as everywhere taming and improving the African wilderness. Several photographs depicting European settlers incorporating wild "pets" into the domestic sphere of the estates accompany his narrative.[36] These photos and the many others of carcasses of wild beasts drive home the message of domination of Europeans over African nature.

Roosevelt further arranged Africans within a hierarchical system that typically portrayed them as either animal-like or childlike, depending on their rank. Ndorobo, East African hunter-gatherers, occupied the bottom of the scale. Roosevelt associated them closest to primates, as when he observed a Colobus monkey in a tree's "topmost branches where only a monkey or a 'Ndorobo could have felt at home." Ndorobo and other tribes could mix to produce "bastard" races: "Kikuyu 'Ndorobo . . . were . . . merely outlying, forest-dwelling members of the lowland tribes." Bastardized races could shift in the hierarchy. "In the deep woods we met one old Dorobo, who had no connection with any more advanced tribe . . . ; unlike the bastard 'Ndorobo, he was ornamented with neither paint nor grease. But the 'Ndorobo who were our guides stood farther up in the social scale." As with all such hierarchies, there were contradictions and ambivalences. Nomadic pastoralists, while more "primitive," were given high praise. "Masai were fine, daring fellows." They "had the erect carriage and fearless bearing that naturally go with a soldierly race," and the "men were tall, finely shaped savages." Once disciplined to life as porters, Africans of all tribes became "patient childlike savages, who have borne the burdens of so many masters and employers hither and thither." Because the porters have "something childlike about them that makes one really fond of them," a spirit of paternal discipline is needed. A master should "grow to feel for them, and to make them in return feel for him, a real and friendly liking." Domesticated through wage labor, porters were better off than their surrounding tribesmen "simply because they were on a white man's safari." The ideal of paternal discipline was extended to the entire region including the Bugandas, who Roosevelt believed represented "a veritable semi-civilization, or advanced barbarism," at the top of the hierarchy. Nurtured by "widely spread rule of a strong European race," "nascent cultures, nascent semi-civilizations," such as the Buganda, would be given the chance to develop (*Game Trails*, 419, 293, 124, 202, 23, 45, 94, 280, 429, and 431).

Signs in the landscape that may have contradicted Roosevelt's imagined savage wilderness were ignored or explained away. After meeting African farmers along the road to Lake Albert carrying "long bales of cotton on the heads," he made no mention of local entanglements in global commodity markets. Finding the remains of stonewalls on the Uasin Gishu plateau, he reads them as evidence of cultural devolution rather than as a challenge to his Pleistocene vision. "They were certainly built by people who were in some respects more advanced than the savage tribes who now dwell in the land"; however, "they have been engulfed in the black oblivion of a lower barbarism." In a rare instance where he acknowledged productive African agrarian landscapes, Roosevelt quickly relegates their creators to the status of caricatured savages. Near Mount Kenya he found that the "cool, shady banana plantations, fenced in with tall hedges and bordered by rapid brooks, were really very attractive. Among them were scattered villages, and . . . grain . . . stored in huts raised on posts. There were herds of cattle, . . . flocks of sheep and goats," and "huge bottles for milk." Nevertheless, the farmers, dressed in their leaf skirts, reminded him of "the pictures of savages in Sunday-school books" rather than the white pioneers of Wyoming and Montana (*Game Trails*, 339, 405, and 311).

Wildlife Tourism: Reenacting the Reenactment

Roosevelt, as I pointed out, traveled a path in East Africa that had already become a well-worn tourist circuit. He himself suggested to his readers that "no place could be more attractive to visitors. There is no more danger to health incident to an ordinary trip to East Africa than there is to an ordinary trip to the Riviera." If he did not exactly invent African wildlife tourism, his public writing certainly propelled the fledgling industry to new heights. At times he wrote as an industry booster, promoting the region as "an ideal playground alike for sportsmen, and for travelers who wish to live in health and comfort, and yet to see what is beautiful and unusual." Even in those early years of the tourism industry we see "traditional" African native practices organized as spectacles for whites. At Uasin Gishu,

Roosevelt described how two white settlers and the district commissioner arranged "for a party of Nandi warriors to come over and show me how they hunted lion." A number of other settlers and the commissioner's wife "had also come to see the sport." In the spirit of the spectacle staged for white amusement, Roosevelt writes in the style of a carnival huckster hyping the crowds. The party finds a lion "and now the maned master of the wilderness, the terror that stalked the night, the grim lord of slaughter, was to meet his doom at the hand of the only foes who dared molest him." The white party cornered the lion on horseback and waited for the warriors, who then surrounded him and killed him with repeated spear thrusts. Two warriors were seriously mauled, adding to the thrill of the spectacle for the white audience. The spectacle complete, the paternal instincts of the civilizing race returned. The party treated the warriors' wounds with antiseptic while Roosevelt promised a tip for the performance. He told them that "I would give each a heifer," and thus "each sufferer smiled broadly at the news, and forgot all about the pain of his wounds." Lest the significance of this spectacle to his adventure narrative be discounted, it is worth noting that five photos and a full-page color illustration (see figure 7) are dedicated to this one staged event, about which Roosevelt doubted "whether ten seconds had elapsed" (*Game Trails*, 148 and 406–10).

What has Roosevelt's writing to do with African wildlife tourism today? His book was named the *New York Times* book of the year, attracted millions of readers, propelled a wave of interest in African travel, and spawned a small industry of imitative texts that crowded the bookstore windows of early twentieth-century New York.[37] Clearly, however, no tourism marketer and few tourists today express a view of Africa in the Late Victorian lexicon that includes "bastard" races, "barbarism," and "pure savages." Nor is any tour package likely to include an African lion hunt with spears. Nevertheless, the widespread popularity of Roosevelt's writing in America did serve to firmly establish certain narrative tropes, racial hierarchies, essentialized identities, and geographic imaginaries that script tourism in Africa today. Depictions of Africa as a timeless wilderness are ubiquitous in contemporary promotional literature of wildlife tourism. If Africa is wilderness, then Africans cannot be fully developed, though they are no longer barbarous, merely "traditional" in their (now "colorful") tribal ways. Moreover, Roosevelt's performance as an adventurer establishes a tourist archetype. Just as he imagined encountering "a wilderness of savage men and beasts" and "being back in the days of Cornwallis Harris and Gordon Cumming," wildlife tourism marketers invite Americans to imagine visiting a land "existing today as it did when Teddy Roosevelt first traveled to Africa in

1909."[38] Nearly every promotion for hunting safaris in East Africa suggests an experience reminiscent of Roosevelt's. Colonial nostalgia and reenactment are explicit in today's advertising. *Africa Travel Magazine* suggests that stopping at Nairobi's Norfolk Hotel "is like taking a journey into British colonial history. . . . The Norfolk was the base from which many great adventurers began [and] where US President Roosevelt began his world famous safari in 1909 setting out from the front steps of the hotel into the wilderness."[39] East African wildlife tourism today is staged and performed using generic, generalized versions of Roosevelt's script, which was in turn based on the scripts of Victorian hunter-naturalists. We are invited to reenact Roosevelt's reenactment and so enter a timeless hall of mirrors in which experience becomes a reflection of an imagined place in an imagined past.

Roosevelt's writing remains salient to the American geographic imaginary today, as evidenced by the events and products commemorating the centennial anniversary of his expedition. To mark the safari's centennial in 2009, Winchester Repeating Arms produced a gun "patterned after one of Roosevelt's Model 1895 rifles." Each limited-production two-rifle set "will bear special matching serial numbers, and is supplied with a complete collector's package, including a . . . copy of President Roosevelt's famous book *African Game Trails.*"[40] At least two presses have produced special reproductions of the book to celebrate "the 100-year anniversary of the greatest of all African hunts by an American president" with "the highest quality reprint ever published of this title." The Theodore Roosevelt Association, Capital Area Chapter, used the occasion of its annual dinner in 2009 to "mark the centennial of Theodore Roosevelt's renowned African Safari" with a presentation by the director of the Smithsonian Institution's Institutional History Division.[41] The Piedmont Club in Winston-Salem brought in Roosevelt's great-grandson Tweed Roosevelt to recount the 1909 safari to generate interest in an exclusive private safari commemorating the anniversary.[42]

Marketed as "The Roosevelt Safari: A Centennial Celebration of President Theodore Roosevelt's 1909–1910 East African Safari," the event that Tweed helped promote represents one of the most explicit and direct reenactments of Roosevelt's adventure. The tour, conducted by the CEO of the Palm Beach Zoo, uses *African Game Trails'* "exciting narrative as a guide" to create "a modern celebration of Roosevelt's pioneering safari." As Roosevelt did, the party would stay at early settler estates, which provide "a glimpse into life in Kenya during much of its colonial era." Though sponsored by a zoo and focused on wildlife viewing, colonial reenactment is a critical part of the package. "At Loldia we step back in time and experience the life of Kenya's early European settlers who cleared the forests and first farmed these rich highlands. Loldia Ranch was established by a family who trekked by ox-wagon from South Africa a century ago and whose descendants still own the land." As with Roosevelt's adventure, Africans are organized as part of the tourist spectacle. No longer bearing the reputation of "a soldierly race" of "finely shaped savages," Maasai are now "[k]nown for their stamina over long distances" and as the "most colorful African tribe." For tour participants "who are runners, a run through the countryside with a Maasai warrior can be an exhilarating experience," though not as exhilarating perhaps as staging Nandi warriors in a lion hunt with spears.[43] Of course, the main attraction will be stalking African wildlife, albeit with cameras rather than Roosevelt's "Big Stick."

Conclusion

The most elaborate and extensive naval parade ever before assembled for a dignitary greeted Roosevelt as he returned to New York Harbor from his African adventure. Three different ships, including the newly commissioned battleship *South Carolina,* fired twenty-one–gun salutes as his luxury steamer passed.[44] Once Roosevelt was on shore, Cornelius Vanderbilt pinned a medal on his chest, the mayor shook Roosevelt's hand, and assembled Rough Riders slapped his back. Roosevelt then was treated to a tickertape parade and cheered by a crowd of one million along a winding five-mile route through Manhattan.[45] In sum, he was feted as a conquering hero or path-breaking explorer rather than recognized as someone returning from a lavish wildlife adventure tour in a British settler colony. Yet it was widely known at the time that if one could afford the steamer passage and outfitter fees, the same adventure was available to anyone. Even the *New York Times* seemed to occasionally be weary of reporting on the year-long safari, offering headlines such as "Roosevelt Hunt Goes On," "More Game Slain," and "Roosevelt Still Shooting." Moreover, collecting expeditions such as Roosevelt's were so common at the time that they bumped into one another. Hunting in the East African highlands, Roosevelt (*Game Trails,* 399) "saw a white man in the trail ahead" who turned out to be "Carl Akeley, who was out on a trip for the American Museum of Natural History in New York." Later while hunting near Mount Meru, Roosevelt's expedition "encountered the safari of an old friend, William Lord Smith," who was "on a trip that was . . . partly a scientific trip taken on behalf of the Cambridge Museum" (404). What, then, elevated Roosevelt's trip to the level of Peary's and Shackleton's expeditions in mass print media and the public imagination and warranted a tickertape parade?

Certainly Roosevelt's celebrity and political status and "prodigious" "skill as a self-advertiser" explain much of the excitement and hype.[46] Another part of the answer is found in Said's concept of imagined geographies and the silences and selectivity that are critical to discursive construction of the geographies of the Other. Roosevelt did not invent the idea of Africa as Pleistocene wilderness; he adopted it from numerous pre-texts and elaborated on it for the entertainment of his readers, the elevation of his own accomplishments, and the fulfillment of his adventure. Wild and savage Africa had already been discursively established prior to his departure. To maintain this geographic imaginary, certain details—bales of cotton, well-tended African fields and villages, the garbage from previous tours—had to be ignored or explained away. Travel writers and their readers must engage in a willing suspension of disbelief and imagine themselves experiencing a landscape of a distant time and place: Africa frozen in the Pleistocene. The power of representational discourse lies in erasing not only alternative representations but also observable physical and historical facts.

As travel literature, Roosevelt's writing is unremarkable. He recycles all of the banal characteristics of the genre—the nostalgia, the categorization of observations in terms of established tropes, the cultural superiority of the white traveler—that were shopworn even then. Nevertheless, the didactic importance of his writing and its role in the discursive construction of African nature and race cannot be overstated. As Schlesinger observed, "Roosevelt transfixed the imagination of the American middle class as did no other figure of the time."[47] Roosevelt's African safari held the nation's attention for more than a year, and his book continues to embody the African adventure experience for new generations of readers. Today's readers may wince at the antiquated terminology used to construct racial hierarchies, though perhaps less so at the celebration of white accomplishment and positioning of Africa as perpetually backward. As geographer Allan Pred reminded, "Once representations of the racialized Other have taken on a life of their own, once they have become widely disseminated through discursive networks, . . . those representations . . . become a condition of the minority Other's everyday practical and political existence."[48] Even for the many wildlife tourists who have never read *African Game Trails*, Roosevelt's African experience continues to resonate. It is clear from the advertising copy cited here that tour operators believe that references to Roosevelt and Anglo-American empire will evoke a set of feelings and associations about Africa, specifically as a place stuck in a prehistoric past awaiting the great white explorer. Through such representations, Roosevelt instructs us even now on African nature and "race" and positions us in relation to both.

Notes

All illustrations are courtesy of the Theodore Roosevelt Collection, Harvard College Library. I wish to thank Wallace Daily, the curator of the collection, for his generous advice and assistance. Thanks are also due to the staff of the Zimbabwe National Archives for allowing me access to the Selous-Roosevelt correspondence on very short notice. A special thanks is due to my colleague and spouse, Gail Hollander, whose love of old books and me brought *African Game Trails* into my possession. The comments of two anonymous reviewers were helpful in revising the original manuscript. Finally, I am grateful to Garth Myers and Byron Caminero-Santangelo for the chance to contribute to this project.

1. "Roosevelt Happy on Eve of Sailing," *New York Times,* March 22, 1909. All *New York Times* articles were accessed at http://query.nytimes.com/mem/archive-free/ on January 24, 2009.

2. "Roosevelt's Outfit Ready in Africa," *New York Times,* March 28, 1909.

3. "If Citizen Roosevelt Should Tour Europe," *New York Times,* September 6, 1908; "Danger to Roosevelt," *New York Times,* March 7, 1909; and Winston Churchill, "Along the Line of Big Game in Africa," *New York Times,* May 16, 1909.

4. "Roosevelt Happy on Eve of Sailing," *New York Times,* March 22, 1909; "Roosevelt on View as He Sails Away," *New York Times,* March 23, 1909; and "Roosevelt Sails in Roar of Cheers," *New York Times,* March 24, 1909.

5. Francis Collins, "Mr. Roosevelt's Africa," *New York Times,* August 27, 1910.

6. The *New York Times* reported that "After his return in the Roosevelt from the discovery of the North Pole, Commander Peary delivered his initial lecture to the society, and Sir Ernest Shackleton's first address in America on his Antarctic explorations was under the same auspices. It is thought probable, therefore, that Mr. Roosevelt will honor the society in a similar manner." "Roosevelt Also to Lecture," *New York Times,* April 10, 1910.

7. For this chapter I have consulted two archives containing Theodore Roosevelt's personal papers. The Zimbabwe National Archives in Harare contains his correspondence with Frederick Courtney Selous. The Theodore Roosevelt Collection in the Houghton Library at Harvard University contains a variety of his personal records, including correspondence and memorabilia from his African safari. I also consulted the *New York Times* online archive at http://query.nytimes.com/mem/archive-free/. For present-day tourism

advertisement I analyzed a variety of online and print materials produced by travel businesses.

8. Eric Hobsbawm, *The Age of Empire, 1875–1914* (New York: Vintage Books, 1987).

9. Mona Domosh, *American Commodities in the Age of Empire* (New York: Routledge, 2006).

10. Susan Schulten, *The Geographical Imagination in America, 1880–1950* (Chicago: University of Chicago Press, 2001), 39.

11. Mary Louise Pratt, *Imperial Eyes: Travel Writing and Transculturation* (London and New York: Routledge, 1992), 5, emphasis in original.

12. Edward Said, *Orientalism* (London: Routledge and Kegan Paul 1978); Derek Gregory, "Between the Book and the Lamp: Imaginative Geographies of Egypt, 1949–50," *Transactions of the Institute of British Geographers,* n.s. 20, no. 1 (1995): 29–57; Sara Mills, "Knowledge, Gender, and Empire," in *Writing Women and Space, Colonial and Postcolonial Geographies,* edited by Alison Blunt and Gillian Rose (New York: Guilford, 1994), 29–50; James Duncan and Derek Gregory, eds., *Writes of Passage: Reading Travel Writing* (London: Routledge 1999); and Garth Myers, "Colonial Geography and Masculinity in Eric Dutton's Kenya Mountain," *Gender, Place and Culture* 9, no. 1 (2002): 23–38.

13. John Gascoigne, "The Expanding Historiography of British Imperialism," *Historical Journal* 49, no. 2 (2006): 578.

14. Duncan and Gregory, *Writes of Passage,* 3.

15. Derek Gregory, "Between the Book and the Lamp: Imaginative Geographies of Egypt, 1949–50," *Transactions of the Institute of British Geographers,* n.s. 20, no. 1 (1995): 29.

16. Derek Gregory, "Scripting Egypt: Orientalism and the Cultures of Travel," in *Writes of Passage,* 116.

17. In addition to Pratt, *Imperial Eyes,* and Duncan and Gregory, *Writes of Passage,* see Bruce Braun, "Colonialism's Afterlife: Vision and Visuality on the Northwest Coast," *Cultural Geographies* 9 (2006): 577–92.

18. The half of this correspondence originating from Roosevelt is kept in the Zimbabwe National Archives, SE 1/1/3 (hereafter ZNA).

19. Theodore Roosevelt to Frederick Selous, May 25, 1907, ZNA.

20. Theodore Roosevelt to Frederick Selous, June 25, 1908, ZNA.

21. Theodore Roosevelt to Frederick Selous, April 29, 1908, and June 25, 1908, ZNA.

22. Theodore Roosevelt, *The Wilderness Hunter* (New York: J. P. Putnam's Sons, 1893), 11, and Theodore Roosevelt to Frederick Selous, September 12, 1908, ZNA.

23. Theodore Roosevelt to Frederick Selous, August 19, 1908, ZNA.

24. Theodore Roosevelt to Frederick Selous, March 19, 1900, ZNA.

25. "Helped to Finance Roosevelt's Hunt," *New York Times,* February 17, 1913.

26. Roosevelt wrote to his wife during the trip that "If things go on as they have begun, I shall turn up at Khartoum with the year's trip having cost me only one half of what Scribner pays me." Theodore Roosevelt to Edith Roosevelt, August 1, 1909, Juja Farm, British East Africa, File bMS Am 1834 (822–57), Theodore Roosevelt Collection, Houghton Library, Harvard College (hereafter TRC).

27. "Roosevelt's Outfit Ready," *New York Times,* March 28, 1909.

28. M. Wright, "East Africa 1870–1905," in *The Cambridge History of Africa,* Vol. 6, edited by R. Oliver and G. N. Sanderson (Cambridge: Cambridge University Press), 539–91, and Harry Johnston, *A History of the Colonization of Africa by Alien Races* (New York: Cooper Square Publishers, [1913] 1966).

29. Winston Churchill, *My African Journey* (New York: Norton, [1908] 1989), 49.

30. "Woman Went over Roosevelt's Route," *New York Times,* March 13, 1909.

31. For example, going after lion from Pease's house, Roosevelt (81) was joined by a "fellow-guest, Medlicott, and not only our host, but our hostess and her daughter; and we were joined by Pecival at lunch."

32. Theodore Roosevelt to Edith Roosevelt, September 2, 1909, File bMS Am 1834 (822–57), TRC.

33. Theodore Roosevelt to Anna Roosevelt Cowles, January 21, 1910, in the Lado Enclave, File T. R. Correspondence Photostats, Box 38, TRC.

34. A few settlers and the district commissioner arranged for Nandi warriors to use spears to kill a lion for Roosevelt's entertainment.

35. The idea that Victorian obsessions with classification of nature, racial hierarchies, nature conservation, and racial purity were all of the same project has been fruitfully explored, albeit from very different theoretical positions, by John MacKenzie, *The Empire of Nature: Hunting, Conservation, and British Imperialism* (Manchester, UK: Manchester University Press, 1988), and Donna Haraway, "Teddy Bear Patriarchy: Taxidermy in the Garden of Eden, New York City, 1908–1936," *Social Text* 12 (1984): 20–63.

36. Juja Farm, for example, had a leopard, five lions, three cheetahs, a warthog, and a Grant's gazelle as "pets." In one photo Mrs. McMillan, in full English Victorian dress, walks a cheetah on a leash (147).

37. "New York Book Announcements," *New York Times,* June 25, 1910.

38. Ibid., 1 63, and King's Way Travel, http://kingswaytravel.com/2.htm (accessed April 3, 2009).

39. *Africa Travel Magazine,* http://www.africa-ata.org/ke_norfolk.htm (accessed April 3, 2009).

40. Winchester Repeating Arms, http://www.winchesterguns.com (accessed April 3, 2009).

41. Theodore Roosevelt Association, Capital Area Chapter, http://www.trassociation.org/mc/community/eventdetails.do?eventId=213866&orgId=tra (accessed April 3, 2009).

42. "Teddy Roosevelt's Great-Grandson Will Discuss Former President's African Safari," *Business Journal,* August 12, 2008, www.bizjournals.com/triad/stories/2008/08/11 (accessed April 3, 2009).

43. All quotes are from "The Roosevelt Safari: A Centennial Celebration of President Theodore Roosevelt's 1909–1910 East African Safari," http://www.palmbeachzoo.org/pdf/travel/2009-travel-program-roosevelt.pdf (accessed April 3, 2009).

44. "Million Join in Welcome to Roosevelt," *New York Times,* June 19, 1910.

45. Ibid.

46. "Mr. Roosevelt's Return," *New York Times,* June 18, 1910.

47. Arthur Schlesinger Jr., *The Crisis of the Old Order, 1919–1933* (Boston: Houghton Mifflin, 1957), 18.

48. Allan Pred, *The Past Is Not Dead: Facts, Fictions and Enduring Racial Stereotypes* (Minneapolis: University of Minnesota Press, 2004), 6.

Bibliography

Braun, Bruce. "Colonialism's Afterlife: Vision and Visuality on the Northwest Coast." *Cultural Geographies* 9 (2006): 577–92.

Churchill, Winston. *My African Journey.* New York: Norton, [1908] 1989.

Domosh, Mona. *American Commodities in the Age of Empire.* New York: Routledge, 2006.

Duncan, James, and Derek Gregory, eds. *Writes of Passage: Reading Travel Writing.* London: Routledge, 1999.

Gascoigne, John. "The Expanding Historiography of British Imperialism." *Historical Journal* 49, no. 2 (2006): 577–92.

Gregory, Derek. "Between the Book and the Lamp: Imaginative Geographies of Egypt, 1949–50." *Transactions of the Institute of British Geographers,* n.s. 20, no. 1 (1995): 29–57.

———. "Scripting Egypt: Orientalism and the Cultures of Travel." In *Writes of Passage: Reading Travel Writing,* edited by James Duncan and Derek Gregory, 114–50. London: Routledge, 1999.

Haraway, Donna. "Teddy Bear Patriarchy: Taxidermy in the Garden of Eden, New York City, 1908–1936." *Social Text* 11 (Winter 1984–85): 20–63.

Hobsbawm, Eric. *The Age of Empire, 1875–1914.* New York: Vintage Books, 1987.

Johnston, Harry H. *A History of the Colonization of Africa by Alien Races.* New York: Cooper Square Publishers, [1913] 1966.

MacKenzie, John. *The Empire of Nature: Hunting, Conservation, and British Imperialism.* Manchester, UK: Manchester University Press, 1988.

Mills, Sara. "Knowledge, Gender, and Empire." In *Writing Women and Space: Colonial and Postcolonial Geographies,* edited by Allison Blunt and Gillian Rose, 29–50. New York: Guilford, 1994.

Myers, Garth. "Colonial Geography and Masculinity in Eric Dutton's Kenya Mountain." *Gender, Place and Culture* 9, no. 1 (2002): 23–38.

Pratt, Mary Louise. *Imperial Eyes: Travel Writing and Transculturation.* London and New York: Routledge, 1992.

Pred, Allan. *The Past Is Not Dead: Facts, Fictions and Enduring Racial Stereotypes.* Minneapolis: University of Minnesota Press, 2004.

Roosevelt, Theodore. *African Game Trails: An Account of the African Wanderings of an American Hunter-Naturalist.* New York: Charles Scribners' Sons, 1910.

Said, Edward. *Orientalism.* London: Routledge and Kegan Paul, 1978.

Schlesinger, Arthur, Jr. *The Crisis of the Old Order, 1919–1933.* Boston: Houghton Mifflin, 1957.

Schulten, Susan. *The Geographical Imagination in America, 1880–1950.* Chicago: University of Chicago Press, 2001.

Wright, Marcia. "East Africa 1870–1905." In *The Cambridge History of Africa,* Volume 6, edited by R. Oliver and G. N. Sanderson, 539–91. Cambridge: Cambridge University Press, 1974.

Chapter 3

"Hunter of Elephants, Take Your Bow!"

A Historical Analysis of Nonfiction Writing about Elephant Hunting in Southern Africa

Jane Carruthers

THE PURPOSE of environmental history is to probe the nexus between humans and nature (the environment). Like all historical studies, environmental history relies on the critical evaluation of sources, usually but not exclusively the written word. Moran asserts that literary studies and history have had a close but problematic relationship, and this may well be evident in what follows.[1] This chapter is based on a selection of nonfiction relating to elephants in southern Africa produced over the course of more than a century.[2] Considering how hunting, killing, and managing elephants has been expressed in nonfiction elucidates changing attitudes toward this species, the environment, and ecopolitics and demonstrates transformations within our constructions of nature and culture. Elephants seem to be particularly appropriate subjects for such an analysis because they are the largest of the charismatic African land mammals and bear a special burden of cultural constructs and ethics.[3] As arguments about the future of elephants clearly show, people are polarized, debates around conflicting value systems are sharpened, and emotions are highly charged.

Discussing African elephants in terms of current environmental histo-riography is also appropriate. The "animal turn" in the social sciences has proliferated in the field of ecocriticism,[4] although a good deal of this litera-ture is avowedly political, even polemic.[5] Nonetheless, Harriet Ritvo argues that there is currently more "animal history" because of the political pur-chase and high profile of animal-related causes as well as the growth of en-vironmental history.[6] Both Ritvo and Keith Thomas have pointed out that the history of the treatment of animals tells us a great deal about human societies.[7] Thomas (*Man and the Natural World,* 166) identified the con-texts in which an anthropocentric tradition was eroded, while John Mac-Kenzie gave the natural world a dominant role in the imperial enterprise.[8]

Considering elephants provides a rich gateway into aspects of human thinking, especially in connection with current changes in interspecies eth-ics.[9] Elephants, together with a number of other animal genera, are spear-heading the increasingly vocal animal rights movement, a development that historian and wilderness advocate Roderick Nash foresaw many years ago. On the basis of an ever-extending network of rights that he suggested had incrementally included slaves, women, and indigenous people, Nash argued that animals would be "liberated," and eventually so would the en-vironment itself.[10]

In some respects these debates can be encapsulated by considering the growing political relevance of nonhuman animals (and the radicalization of the animal rights movement) as well as the contrasting philosophies of sustainable use and the primacy of any animal's right to life. Moreover, Jamieson has referred to the opposition between "animal values" and the "value of nature" (or ecocentrism). Critics of the animal rights ethical framework have referred to its ideas as equal to anthropomorphism, that is, as similar to the obsession with humans.[11]

African and Asian Elephants: The Wild and the Tame

In past centuries African elephants were used for entertainment in Roman games and circuses and played their part in warfare. There are also isolated records of African elephants being tamed as beasts of burden, and today some captive elephants are used for safari treks. However, they have never been domesticated to the extent that they have in India and elsewhere in South Asia. Asian elephants are tamed, are helpers in a number of work environments, have status as domestic stock, and are economically useful. In his powerful anti-imperial essay *Shooting an Elephant,* George Orwell expressed something of their utilitarian value. After an elephant, a male in musth, had killed a man, Orwell was obliged to shoot the animal. Orwell lamented that "Alive, the elephant was worth at least a hundred pounds; dead, he would only be worth the value of his tusks, five pounds, possibly."[12]

What a contrast there is between this Burmese elephant and Africa's herds! Unlike the Asian elephant of which little is heard globally today, Africa's elephants are a topical subject because they are worth so much more dead than alive. It is their ivory that has international value, and there seems to be no diminution in how avidly it is sought after. Although mostly fenced into national parks and other protected areas, Africa's elephants are wild, not domesticated. In many parts of Africa they are illegally hunted, and their numbers are dwindling to the point of real concern. As is well known, there is an international ban on the ivory trade so that elephants in East and West Africa can recover their numbers. In southern Africa this debate is intense for the opposite reason. There are thought to be "too many" elephants, and they are having an adverse effect adversely on many environments. Humans have intervened to decide what to do about limiting the increasing numbers and minimizing the habitat destruction that an overpopulation causes. On the one hand, there is a powerful argument for constraining elephant populations in terms of habitat damage and human-elephant conflict, and in this regard initiatives such as culling, translocation, and contraception are being considered.[13] On the other hand, the argument for sustainable use is influential in the African context because money raised from selling ivory might provide rural clinics, schools, roads, and other infrastructure that governments in southern Africa battle to supply to their citizens, hobbled as they are by the paucity of resources in the crisis of the HIV/AIDS pandemic and the challenges of underdevelopment and lack of capacity.

Elephants in Africa

So geographers, in Afric maps,
With savage-pictures fill their gaps;
And o'er uninhabitable downs
Place elephants for want of towns

(Jonathan Swift, *On Poetry: A Rhapsody*, lines 174–77)

These well-known lines by Jonathan Swift (1667–1745) are often quoted and dismissed as a satire on the inventive cartography of the age. However, they contain three significant observations about the elephant populations of Africa that illuminate aspects of elephant distribution and human-elephant contact and continue to influence elephant management. The first is that elephants are the iconic and most charismatic mammals of Africa, indeed its very symbol. As Adams and McShane observe, "No other species carries as much symbolic or emotional force. Fascination with elephants is hardly a new phenomenon—man has by turns worshipped, idealized, contemplated, or slaughtered elephants, but rarely ignored them."[14] The second observation is that elephants were once widely distributed on the African continent, occurring wherever there was suitable habitat, while the third observation is that where large settled concentrations of humans occur, there are either no elephants or very few.

Elephants—highly intelligent animals—have played a crucial role in Africa's history. Ivory is by far their most significant by-product. Unlike numerous other natural resources, ivory does not deteriorate quickly, humans can transport it, and over the centuries it has retained its high commercial value.[15] Modern Western writing about elephants generally does not extol the beauty of ivory. At a time when avarice for this product is the reason why many influential people in the Western world mourn the diminution of the elephants, to harp on its aesthetics and the satisfaction of ownership might be regarded as irresponsible, let alone politically incorrect. Nonetheless, for more than ten thousand years the "subtle glowing colour and sensual surface" of ivory has ensured its prominent position among the world's luxury goods.[16]

The ivory trade is particularly sensitive to taste and fashion. A major peak in supply and demand occurred during the nineteenth century with the industrialization of Europe and the United States. At that time, practical as well as luxury objects made of ivory became very popular with the growing middle class, and the market expanded. In the early decades of the twentieth century, ivory was used less frequently in the West, but in the 1970s demand from Asia peaked, and this took its toll particularly in East Africa, leading to a ban on ivory exports from Africa.[17]

The power base of many African leaders lay in their ability to control the ivory trade, and extraction of tribute enabled Africans to dominate and exploit each other. Access to ivory laid the foundations of strong states and the emergence of hierarchies based on wealth and class. But ivory extraction has never promoted a strong and sustainable economy and instead has resulted in a fragile economy based on a single product.

Nineteenth-Century Hunters

Against this background of economic wealth through ivory hunting we need to thread the hunting narrative. A strong trope in nineteenth-century literature in English about Africa is the "hunting journal"[18] and, as indicated in Neumann's account (in the previous chapter) of Theodore Roosevelt's well-orchestrated and publicized East Africa safari, also into the twentieth-century North American canon. When Britain took control of Cape Colony in 1806 at the end of the Napoleonic Wars, colonial settlement expanded. Pioneer trekboers and traders of the late 1700s and early 1800s who penetrated the interior seeking ivory and other commercial products of the hunt were followed by, for example, the scientific expeditions of Andrew Smith and recreational sport hunters such as William Cornwallis Harris and Roualeyn Gordon Cumming. More significant in terms of numbers were the large Voortrekker parties who were not visitors and itinerants but potential settlers cementing partnerships with African mercenary hunters and seeking to establish independent polities.[19] Due to the belief that ridding the countryside of wildlife was a patriotic necessity in order to create a "civilized" state, elephant hunting was a major factor accelerating colonial expansion into the interior of southern Africa, and the hunter-trader lifestyle was encouraged by settler society.[20] This period saw a shift in the power relations between wildlife (particularly elephants and lions) and humans as firearms spread throughout the subcontinent, with humans gaining the upper hand.

The settler commercial hunters—generally Boers—have not bequeathed a literature detailing their exploits, and what we know of their activities comes from evidence in formal debates around the law and from observations about them from literate travelers. One of the most famous of the memoir-writing sport-hunter travelers was also the first. William Cornwallis Harris, whose books initiated a new genre in African literature and continued to exert an influence on later hunter-visitors such as Theodore Roosevelt (see Neumann, this volume). Harris, a British Army officer stationed in India, arrived in South Africa in 1836 to spend his leave on a hunting expedition. He devoted a considerable part of his subsequent account—an extremely popular adventure book, illustrated with his own excellent depictions of landscape and animals—to describing how he hunted elephants. The ivory that he amassed funded his expedition, but he also hunted for entertainment and for what he regarded as "science." He captures the drama well. An elephant footprint, imprinted during a particularly fierce thunderstorm, is carefully measured, and according to Harris (*Wild Sports*, 167), because science has determined that the height of an elephant is twice the circumference of the footprint, he has the spoor of the largest specimen possible, twelve feet tall. A scene in the Magaliesberg—a range of mountains straddling Pretoria and Rustenburg—of three hundred elephants is, for Harris, a "grand panorama," "a picture at once soul-stirring and sublime." From the description that he offers of this incident, it is clear that Harris is among a herd of females (which he refers to as "ladies") and their young. After he has "attacked" and killed some of the females, he realizes that the young are deeply attached to their mothers, one of them going around a corpse "with touching demonstrations of grief, piping sorrowfully and vainly attempting to raise her with its tiny trunk." Harris later confessed that "I had felt compunctions in committing the murder the day before, and now half resolved never to assist in another," an emotion strengthened because he could not rid himself of the feeling that he was killing his own tame elephant at home in India.[21] Despite his ethical sensitivity that what he was doing might be construed as "murder," Harris did not desist, and his book abounds with descriptions of profligate killing and details of the gory process of extracting tusks. Sales were substantial, and Harris's exciting adventures and his proficient illustrations introduced many European readers to the landscape of southern Africa and to its wildlife and people. Harris hunted in southern Africa at a time when European imperialism had not yet taken a grip in the region, and he thus gloried in his freedom from any authority. His appreciation of a landscape appropriately filled with elephants shows his familiarity with "the picturesque," and he knew that his illustrations would appeal to his readers. The "delight" in elephant slaughter harks back to a long hunting tradition that provided entertainment and recreation for the nobility in Europe and also to then-current ideas about masculine bravery. However, the articulation of the idea that acting thus might be "murder" indicates, even if dimly, that people such as Harris were cognizant that some nonhuman animals had a right to life and even to family life.

Another example of the hunting genre in this period is the well-known and often-quoted *A Hunter's Life in South Africa* by R. Gordon Cumming, first published in 1850. Cumming (178–80) longed for an encounter with "the noble elephants," but when he got his wish, he killed them at night and tortured them in an experiment to see how long it would take them to die (227–28). The killing gave him economic and social power: he exulted in being able to enrich himself through ivory sales and also to feed the hundreds of local people who followed his wagons and assisted him in hunting (187). Indeed, Cumming (230) summarized his activities by explaining that the killing of an elephant was "so overpoweringly exciting that it almost takes a man's breath away." Nevertheless, he too was not insensitive to the fact that the elephant was a "wonderful animal," that elephants had a defined social structure and patterns of behavior, and that they had a well-developed sense of smell and a means of communicating with one another (180).

The Frenchman Adulphe Delegorgue, whose journals were written in the late 1830s and early 1840s, "took possession" of a dead elephant by walking on its carcass, having given it an extra bullet to stop it from uttering "a sobbing noise" as it lay dying.[22] Frederick Courteney Selous explained in 1881 that "In South Central Africa, at the hunter's camp fire, the elephant takes the place of the grisly 'bar' in North America . . . and there are more yarns spun concerning him than about any other animal." These "yarns," it seems, included horrible details of the killing and dissection process and the methods of extracting tusks.[23]

In addition to the distaste that sensitive modern readers have with the profligate and senseless deaths perpetrated by these men, it is also difficult to reconcile statements such as the following with the actions of the people who made them. Take Delegorgue, for example: "What paltry reason can justify the death and destruction of such beautiful, strong and excellent animals? What are a couple of hundred pounds of ivory compared with the long service which such animals might render to man for generations? . . . I was perfectly conscious of the mischief I was doing but I was a hunter first and foremost. The elephant is reckoned the *ne plus ultra* where *inyamazane* [game] is concerned. I desired no other; all the animals of creation, whatsoever they may be, are as nothing compared with the elephant." As for a young elephant, only about six feet high, Delegorgue (*Travels*, 3–4) wrote that "his face was so comical that I wanted to burst out laughing, for, protruding beyond his lips were tusks only ten inches long. I had a sudden desire to shoot him so that I could inspect him at leisure when he was dead."

These hunters published their adventures at a time when attitudes toward wildlife were changing in Europe. Shortly after he wrote *A Hunter's Life*, Cumming was described as "bold, enterprising and skilful" but a decade or so later was relegated to being "an unprincipled man and an indiscriminate slaughterer."[24] Taking pleasure in killing animals and in causing unnecessary suffering—not necessarily the act of killing itself—became increasingly morally indefensible. James Stevenson-Hamilton, for example, who was on a visit to East Africa at the same time as Roosevelt (and who met the ex-president on that occasion), was one of the new generation of conservationists critical of adventure tourists and distrustful of their motives as nature or wildlife lovers. Stevenson-Hamilton castigated the many "lady explorers in Africa for self-advertisement" and those who demanded hospitality from the local people while they were "tramping about Africa for a bet or notoriety." He had no time for the likes of American millionaire and Roosevelt host W. N. Macmillan and, in contrast to Roosevelt's high opinion of East African settlers as proud standard-bearers of imperialism (as articulated by Neumann, this volume), thought that Nairobi colonists were "a selfish lot, only exploiting" and giving nothing in return. Stevenson-Hamilton's summation was that it was "dreadful to think of the greed and selfishness of low class whites which will cause extermination [of wildlife]—and they are British!"[25]

But the physical and emotional attraction of big-game hunting has endured through subsequent years, and we might speculate that elements of a preindustrial human's fear of wildlife and the excitement of conquest remain strong in some people. Perhaps a later writer of the twentieth century, Robert Henriques, in his book *Death by Moonlight* (1938), best expressed the visceral emotions of hunting elephant: "I hunt big game frankly and brutally because not otherwise can I find the same disappointments and reverses, the same excitements, the same antagonists, and the same exhilarating rewards.... Above all, I get from it a moment of triumph and of pure emotion which I have never found elsewhere."[26]

In his as yet unsurpassed survey of southern African writing, Stephen Gray raised southern African hunting adventures into a literary genre. He observed how the wilderness of southern Africa "becomes a playground for inflated bullies who measure fun in terms of the humbling of brutes greater in size than themselves," where they were able to indulge in sadistic activities. Gray (*Southern African Literature*, 103–4) also identified the mythic element in these texts, particularly the depiction of life as a contest, in which evil, cunning, and intelligent large animals are defeated by humans, and he appreciated too the literary craftsmanship that went into some of the romantic descriptions, the drama of the episodes, and the erudite references to poetry and novels.[27]

Under extreme hunting pressure, the number of elephants in southern Africa decreased from an estimated one hundred thousand before white settlement in 1652[28] to near extermination by 1900. They were reduced to four relict populations totaling fewer than two hundred individuals, located in the forested Knysna area of the Western Cape coastline (30–50), in the dense succulent thicket Addo area of the Eastern Cape (130–40), and an unknown but very small number in the tropical coastal Tembe area of Maputaland, in northern KwaZulu-Natal, and in the Mpumalanga lowveld, the Olifants River gorge on the boundary between South Africa and Mozambique.[29]

A Change in Pace and Discourse in the Twentieth Century

The largest remaining population of the four isolated communities of elephants was in the Addo area of the Eastern Cape.[30] Small in stature with insignificant tusks, these elephants had survived because of the specific local vegetation, which is generally spiky and thorny and is often almost impenetrable. After World War I there were increased initiatives to modernize the economy in this district. Along the Sundays River, close to the Addo bush, the government supported a large and expensive irrigation scheme to encourage the development of large citrus estates. For the elephants this was an irresistible attraction, with water and food in quantity, but was also a dilemma, as "progress" and "wildlife" clashed and farmers complained bitterly about their losses. After a formal state enquiry, the decision was made to kill all of these elephants, this despite the opinion of some people, such as F. W. Fitzsimons, director of the Port Elizabeth Museum, that "the deliberate extermination of these elephants, would, upon grounds of deeply-felt general sentiment, and in the interests of science, be received by not only very high and influential circles in South Africa, but by the general feeling of the civilized world with condemnation, as a step reflecting no credit upon South Africa."[31] Such concerns were of no avail, and the government seems to have had no hesitation in being branded "uncivilized" or of being widely condemned.

The task of extirpating the elephants was delegated to a small dark-skinned man, Phillip Jacobus Pretorius, an Afrikaner and descendant of the Boer hero Andries Pretorius who had led an extraordinarily adventurous life. Phillip Pretorius's autobiography appeared in 1948 as *Jungle Man: The Autobiography of Major P. J. Pretorius C.M.G. D.S.O. and Bar.* The book was not published in South Africa but instead was published in England and Australia and was later translated into Dutch and published in Amsterdam.[32] In chapter 1 of this volume, Garth Myers analyses Eric Dutton's *The Basuto of Basutoland,* a work of nonfiction emanating from the same interwar era as Pretorius's. South Africa was by then, of course, not a British colony but rather a nation-state, and the distinction between Dutton's descriptions of soil erosion and the absence of agricultural modernity in Lesotho may have—at some level—influenced both Pretorius and his government against allowing the "primitive" anachronism of an elephant herd in a developing district of an economically modern and fully independent country to continue.

It is noteworthy that *Jungle Man,* ostensibly conceived in the literary genre pioneered by Harris and Cumming, was the first book in this vein written by a local Afrikaner, and this almost a century after their books. The laudatory foreword was written by Jan Smuts. At the time of the elephant killing Smuts was prime minister of South Africa and, somewhat ironically in terms of his praise of the book and of Pretorius personally, was deeply committed to wildlife conservation and involved in negotiations to establish the Kruger National Park. Smuts wrote that "I have never seen a more thrilling story of a hunter's life. It is full of almost unbelievable incidents, of reckless daring, and of hairbreadth escapes." In big-game hunting, according to Smuts (*Jungle Man,* 5–6), Pretorius "seems to have found his best self-expression and to be the true artist."

Pretorius's book of his Addo exploits is certainly unusual for its time because of the businesslike attitude that the author displays. He describes unusual techniques such as using ladders to get above the scrub in order to see where the elephants were located. Pretorius explained how he had to kill the elephants at exceptionally close range, often in danger of his own life as they charged him from the thickets. He also had to shoot the animals in places, such as along the spine, that even the blood-thirsty Victorian hunters would have regarded as "unsporting." In the course of just over a year Pretorius reduced the number of elephants in the Addo bush from about 130–150 to 16. He boasted of his success: "It was a dramatic thought that I had fought and defeated a family of elephants that had held undisputed sway in that bush for thousands of years" (*Jungle Man,* 210). Unlike the entertainment and ivory that elephant hunting had offered the sport-hunters, Pretorius was obliterating elephants merely because they hampered economic "progress" in a commercial culling operation for which he was well paid by the Cape provincial administration.

Somewhat surprisingly for reasons that are not entirely clear but almost certainly included Pretorius's own wish,[33] as soon as the numbers of elephant had been so drastically reduced, the survivors were given protection within a game reserve, proclaimed by the Cape Province and upgraded in 1931 to the Addo Elephant National Park. Pretorius's book is novel in the discourse around elephant hunting in South Africa, for his was not a metaphorical "military campaign" in which masculine heroes might display their prowess against elephants but instead was a real military campaign. The idea was to vanquish and exterminate an enemy to reduce their numbers and their threat to human welfare. Then too, unlike Harris or Cumming, Pretorius was a paid government official. Hunting the Addo elephant took place only to reduce elephant numbers and to prevent human-elephant contact. In the politics of human-animal relationships this was a reorientation in South Africa, evidence of instrumentalist ethics and of an attitude that regarded wildlife in the "wrong place" or even "too many in the right place" as objects to be obliterated.

The 1960s

By the end of World War II, apart from the government-sponsored extermination of the Addo elephants, there had been a hiatus of nearly fifty years in elephant hunting in South Africa because the numbers had dwindled to the point almost of local extinction. Therefore, by 1950 the South African public, both black and white, had long ceased ivory extraction or elephant hunting.

As mentioned above, at the beginning of the twentieth century the area that was to become the Kruger National Park in 1926 may have contained a couple of individual elephants in the Olifants River gorge on the Mozambique border. However, under conditions of strict preservation on the South African side, their numbers increased, and in addition, elephants migrated along the southern Crocodile River from Mozambique. In 1905 the warden, James Stevenson-Hamilton, estimated that there were 10 elephants in the reserve. By 1925 there were 100 elephants in the reserve, by 1936 there were 250, by 1946 there were 450, and in 1956 there were 1,000. A helicopter count a decade later (1967) gave a confirmed number of 6,586.[34]

The animal turn in South African historiography has generated research in the development of various natural sciences, including the growth of ecology and the evolution of wildlife biology.[35] After World War II, for a number of reasons—including the retirement of the long-standing warden who believed that nature would manage itself, a change of government in 1948, and the increasing influence of ecological considerations rather than species-specific focus—the South African national parks organization (then the National Parks Board) employed a growing number of university-trained scientists who were responsible for managing the vegetation and wild animals in the park along scientific lines. In terms of the prevailing conservation paradigm of that era, the national park was managed in the same way as an extensive cattle ranch, managed—indeed manipulated—for productivity.

This was "scientifically" determined, taking habitat transformation into account. In 1965 the Kruger staff made a decision to maintain the number of elephants in the national park at around seven thousand, the number being determined by the size of the park and the feeding (approximately

270 kilograms per day) and watering (160 liters per day) requirements of elephant. The optimum density was established to be around >0.37 elephants/km^2.[36]

Elephants were culled regularly in the Kruger National Park for almost thirty years. Perhaps not surprisingly, culling did not generate a literary genre but was recorded. The warden's annual reports carried details of the elephants killed every year (as many as 1,846 in 1970 and as few as 16 in 1981, from 1967 to 1997 some 14,629 in total).[37] Family groups were mustered by helicopters, sedated by darting from the air, and killed quickly with well-aimed bullets by teams on the ground. The by-products of the slaughtered elephants were not wasted; their carcasses were loaded onto trucks and taken to an abattoir near (but not within sight of) the Skukuza tourist camp that processed, canned, and dried thousands of culled elephants, buffalos, and hippopotamuses each year.[38] The ivory was extracted and sold until this became illegal.

In the 1990s the East African conservationist Richard Leakey was shown this abattoir and exclaimed that "I watched for a while impressed by the size and scale of the operation, but appalled that this was what wildlife 'management' in the late twentieth century had come to."[39] As far as he was concerned, managing a national park or game reserve had become akin to managing a modern industrial farm, based on an ideology of killing for sustainable utilization and maximum production. In his book *The Last Elephant: An African Quest,* journalist Jeremy Gavron was similarly outraged, describing the Kruger National Park, with its high fences and good roads, as too manicured, almost a Disneyland run on accounting principles (counting numbers) and mass harvesting for profit.[40]

South African nonfiction writers on the Kruger National Park approached their subject differently. One of the most prolific was the journalist Piet Meiring, and among his popular books were *Behind the Scenes in Kruger Park* (1982) and *Kruger Park Saga* (1976). Unlike Gavron, Meiring extolled the Kruger National Park as an example of "untouched nature," "nature's paradise," a place that brings the "public closer to nature." Meiring quoted the then warden of the park, Dr U. de V. Pienaar, as saying that "we want all nature lovers to become part of its blessed atmosphere."[41] Some ten pages later, Meiring (*Behind the Scenes,* 77) seems oblivious to the irony of his support of "the principle of culling and control" to maintain the "correct population balance" in this paradisiacal landscape. He admits (*Behind the Scenes,* 86) that this "necessary" culling is unpopular with the public and that conservationists and game rangers abhor having to kill elephants, but "without such control measures the game reserve would soon degenerate into a desolate wilderness." This is language reminiscent of Nash's *Wilderness and the American Mind:* "If paradise was early man's greatest good, wilderness, as its antipode, was his greatest evil. In one condition the environment, garden-like, ministered to his every desire. In the other it was at best indifferent, frequently dangerous, and always beyond control."[42]

If Leakey was disillusioned to find commercial harvesting of wildlife in South Africa's premier national park, others were delighted to be able to profit from a sustainable and renewable resource. Using the language of science and certainty at a conference on managing mammals in African conservation areas in 1982, Kruger's warden defended the command-and-control model followed in his park as a "pragmatic/economic alternative." Laissez-faire, or nonintervention, as was generally the case in East Africa, Pienaar warned, was "impractical, fraught with danger, inherently untidy and can have unpredictable and even shocking consequences,"[43] presumably referring to what had happened to Tsavo National Park when elephant numbers were unchecked and plummeted with devastating results in a serious drought. He also defended taking manipulative action because modern South Africa could no longer be called really "wild" and conservation areas were relatively small, fenced, and directly influenced by human activity. Water was supplied to wildlife, the savanna grassland was burned regularly, animal populations were monitored and regulated, predators were controlled, and tourists were welcomed, entertained, and educated. This was certainly not an unregulated or "untidy" landscape.

Pienaar delivered his 1982 conference paper with passion and defended his actions logically. As Adams and McShane (*Myth of Wild Africa,* 100) assert, scientists often have the paranoia and passion of writers of "literature." By contrast, however, the paper presented at the same conference by the veterinarians who were actually responsible for culling is chilling because its terminology is so unemotional. Culled animals are described as having been "removed," the humane killing methods are described with detachment, and the abattoir processing is outlined in a technical diagram.[44] This is reminiscent of Cummings's detachment when he described putting bullet after bullet into one of the elephants that he was in the process of killing. To investigate the connection between the culling discourse and apartheid's discourse of this period in South African history is tempting but is outside the scope of this chapter. The connection has, however, recently been made. In a newspaper article in March 2008, a spokesman for Animal Rights Africa in contesting the proposed reintroduction of culling in South Africa is quoted as saying that "The formulation of the government's policy on elephant management, specifically with regard to culling, has been driven by the rampant chauvinistic mindset of utilization that is deeply rooted in colonialism and apartheid, which disregards the inherent value of each individual elephant and commodifies them into unfeeling units purely to be assessed for their recreational and economic value."[45] Others would strongly disagree. The eminent conservationist Richard Bell has argued that until both Africa and nature conservation are divested of fantasy and "romanticism" and acquire an economic value, the illegal killing of animals will continue unabated, a view shared by the philosopher David Schmidtz.[46] Given the continued unsustainable poaching of elephants as occurs in East Africa—dead elephants produce ivory—this argument is not entirely convincing.

Writing in the New Century

In 2001 Martin Meredith published *Elephant Destiny: Biography of an Endangered Species in Africa*. To be sure, elephants are indeed endangered in many parts of Africa but not in southern Africa. In 1900 elephant numbers in southern Africa were close to zero, but it is estimated that today there are around three hundred thousand, some twenty thousand of them in South Africa.[47] The burgeoning number of elephants has become a matter of local and international concern. In 1994 elephant culling ceased abruptly in South Africa's national parks. There were three principal reasons for this decision. First, at the time of the transition to democracy in South Africa that year and with some appreciation that the new government would be strapped for cash to invest in national parks, there was a need to find external sources of funding. This was forthcoming from the International Fund for Animal Welfare (five million rand in 1994) but only if culling was ended. Second, there was insufficient scientific evidence that culling was indeed the best wildlife management method. Third, a new and more flexible management paradigm of adaptive management was being advocated in international conservation biology circles as a counterpoint to command and control. Since culling ended, elephant numbers have doubled in the Kruger National Park, and active public participation and interdisciplinary debate around managing elephants have been in the forefront of South Africa's recent animal politics.

Conclusion

In 2003 Lyall Watson, a biologist with an interest in the supernatural, published *Elephantoms: Tracking the Elephant*. Watson's book is an innovative exploration of elephants in South Africa, like many other nonfiction works about elephants, but is autobiographically based and recounts anecdote and adventure. But Watson avoids becoming embroiled in the culling debate by situating his story along the Cape coast, a specific location in which culling is not an issue because elephants are merely too few, phantoms in fact. The book is worth noting, for Watson suggests that emotion, history, experience, science, and indeed every way of "knowing" are all part of the human-elephant encounter.[48]

In 1979 Stephen Gray (*Southern African Literature,* 111) observed that no southern African work of hunting and adventure had received its reverse image. Despite the plethora of illustrated guidebooks, game ranger reminiscences, and scientific journal articles devoted to southern African mammals, this comment still applies. Nature writing is a large and respected genre in Europe, Australia, and the United States and is mature, sophisticated, and varied. Apparently, however, nature writing is not attractive to the South African reading public. If a literature on nature were to arise, then elephants would be a prime subject. How humanity treats its elephants could, from a literary point of view, become a metaphor for how we humans think about ourselves as a species.[49] Descriptions of how elephants—an intelligent animal that shares many characteristics with humans—have overpopulated the areas to which they are confined to the extent of destroying their habitat and their water and food supply, reducing biodiversity, and behaving "unnaturally," may mask a conversation that humans are having with one another about their own population dynamics and depredations to the planet.

Notes

The quotation in the chapter title is the first line of a traditional African song published in Robert Delort's *The Life and Lore of the Elephant* (London: Thames and Hudson, 1992), 155, and quoted in Dan Wylie, *Elephant* (London: Reaktion, 2008), 90. An earlier version of this chapter has appeared as Jane Carruthers, "Romance, Reverence, Research, Rights: Writing about Elephant Hunting and Management in Southern Africa, c.1830s to 2008," *Koedoe* 52, no. 1 (2010), 6 pages, DOI: 10.4102/koedoe.v52i1.880.

1. Joe Moran, *Interdisciplinarity* (London: Routledge, 2002), 114.

2. For work that deals with fiction, see Wendy Woodward, *The Animal Gaze: Animal Subjectivities in Southern African Narratives* (Johannesburg: Wits University Press, 2008); John A. Murray, ed., *Wild Africa: Three Centuries of Nature Writing from Africa* (New York: Oxford University Press, 1993); and Wylie,

Elephant. This chapter does not explore ideas from books such as ethologist Katie Payne's *Silent Thunder: Hidden Voices of Elephants* (London: Weidenfeld and Nicolson, 1998) or the semifiction of Barbara Gowdy, *The White Bone* (London: HarperCollins, 1999).

3. See H. P. P. Lötter, "Ethical Considerations in Elephant Management," in *Elephant Management: A Scientific Assessment for South Africa,* edited by R. J. Scholes and K. G. Mennell, 370–405 (Johannesburg: Wits University Press, 2008), and Nigel Rothfels, "Elephants, Ethics, and History," in *Elephants and Ethics: Towards a Morality of Coexistence,* edited by Christen Wemmer and Catherine A. Christen (Baltimore: Johns Hopkins University Press, 2008), 101–19.

4. See Greg Garrard, *Ecocriticism* (London: Routledge, 2004).

5. See the Introduction to the present volume.

6. Harriet Ritvo, "History and Animal Studies," *Society and Animals* 10, no. 4 (2002): 403–6. See also Robert Garner, *Animals, Politics and Morality,* 2nd ed. (Manchester, UK: Manchester University Press, 2004).

7. Keith Thomas, *Man and the Natural World: Changing Attitudes in England, 1500–1800* (Harmondsworth, UK: Penguin, 1984); Harriet Ritvo, *The Animal Estate: The English and Other Creatures of the Victorian Age* (Cambridge: Harvard University Press, 1987); Harriet Ritvo, *The Platypus and the Mermaid and Other Figments of the Classifying Imagination* (Cambridge: Harvard University Press, 1997).

8. John M. MacKenzie, *The Empire of Nature: Hunting, Conservation and British Imperialism* (Manchester, UK: Manchester University Press, 1988).

9. See, for example, Christen Wemmer and Catherine A. Christen, eds., *Elephants and Ethics: Towards a Morality of Coexistence* (Baltimore: Johns Hopkins University Press, 2008).

10. Roderick Nash, *The Rights of Nature: A History of Environmental Ethics* (Madison: University of Wisconsin Press, 1989), and Susan J. Armstrong and Richard G. Botzler, eds. *The Animal Ethics Reader* (London: Routledge, 2003).

11. Dale Jamieson, *Ethics and the Environment: An Introduction* (Cambridge: Cambridge University Press, 2008), 143–49.

12. George Orwell, *Shooting an Elephant and Other Essays* (London: Secker and Warburg, 1950), 3–7.

13. See, for example, Scholes and Mennell, *Elephant Management,* and Wemmer and Christen, *Elephants and Ethics.*

14. Jonathan S. Adams and Thomas O. McShane, *The Myth of Wild Africa: Conservation without Illusion* (Berkeley: University of California Press, 1992), 59.

15. R. Luxmoore, "The Ivory Trade," in *The Illustrated Encyclopedia of Elephants,* edited by S. K. Eltringham (London: Salamander, 1991), 148–57.

16. Luxmoore, "The Ivory Trade."

17. Ian Parker and Mohamed Amin, *Ivory Crisis* (London: Chatto and Windus, 1983); Ian Parker, *What I Tell You Three Times Is True: Conservation, Ivory, History and Politics* (Moray, UK: Librario, 2004); and Peter F. Thorbahn, "The Precolonial Ivory Trade of East Africa: Reconstruction of a Human-Elephant Ecosystem" (PhD diss., University of Massachusetts, 1979).

18. Mary Louise Pratt, *Imperial Eyes: Travel Writing and Transculturation* (London: Routledge, 1992); Stephen Gray, *Southern African Literature: An Introduction* (Cape Town: David Philip, 1979); Dorothy Hammond and Alta Jablow, *The Africa That Never Was: Four Centuries of British Writing about Africa* (New York: Twayne, 1970); MacKenzie, *Empire of Nature;* and Ritvo, *The Animal Estate.*

19. William F. Lye, ed., *Andrew Smith's Journal of His Expedition into the Interior of South Africa, 1834–1836* (Cape Town: Balkema, 1975); William Cornwallis Harris, *Portraits of the Game and Wild Animals of Southern Africa* (London: n.p., 1840); William Cornwallis Harris, *The Wild Sports of Southern Africa* (London: Henry Bohm, 1852); and E. J. Carruthers, *Game Protection in the Transvaal, 1846 to 1926* (Pretoria: Government Printer, 1995).

20. Paul Kruger, *The Memoirs of Paul Kruger,* 2 vols (London: Unwin, 1902), 18.

21. Harris, *Wild Sports,* 168–75.

22. Adulphe Delegorgue, *Travels in Southern Africa,* Vol. 1 (Durban: Killie Campbell Africana Library, 1990), 227.

23. Frederick Courteney Selous, *A Hunter's Wanderings in Africa* (Johannesburg: Galago, 1999), 39–47.

24. Jane Carruthers, "Changing Perspectives on Wildlife in Southern Africa, c.1840 to c.1914," *Society and Animals* 13, no. 3 (2005): 189.

25. Jane Carruthers, *Wildlife and Warfare: The Life of James Stevenson-Hamilton* (Pietermaritzburg: University of Natal Press, 2001), 111.

26. Quoted in Hammond and Jablow, *The Africa That Never Was,* 159–60.

27. See also Wylie, *Elephant,* 90–94.

28. Anthony J. Hall-Martin, "Distribution and Status of the African Elephant *Loxodonta africana* in South Africa, 1652–1992," *Koedoe* 35, no. 1 (1992): 65–88.

29. C. J. Roche, "'The Elephants at Knysna' and 'The Knysna Elephants:' From Exploitation to Conservation: Man and Elephants at Knysna 1856–1920," (BA [Hons] diss., University of Cape Town, 1996); C. J. Skead, *Historical Incidence of the Larger Land Mammals in the Broader Eastern Cape,* 2nd ed., edited by A. F. Boshoff, G. I. H. Kerley, and P. H. Lloyd (Port Elizabeth: Centre for African Conservation Ecology, Nelson Mandela Metropolitan University, 2007); Hall-Martin, "Distribution and Status of the African Elephant"; and Ian J. Whyte, "Conservation Management of the Kruger National Park Elephant Population" (PhD diss., University of Pretoria, 2001).

30. M. T. Hoffman, "Major P. J. Pretorius and the Decimation of the Addo Elephant Herd in 1919–1920: Important Reassessments," *Koedoe* 36, no. 2 (1993): 23–44.

31. F. W. Fitzsimons, "The African Elephants," in *The Natural History of South Africa*, Vol. 3, *Mammals* (London: Longmans Green, 1920), 270–71.

32. P. J. Pretorius, *Jungle Man: The Autobiography of Major P. J. Pretorius C.M.G. D.S.O. and Bar* (Sydney: Australasian Publishing, 1948), 187–88.

33. Hoffman, "Major P. J. Pretorius."

34. Ian J. Whyte, Rudi van Aarde, and Stuart L. Pimm, "Kruger's Elephant Population: Its Size and Consequences for Ecosystem Heterogeneity," in *The Kruger Experience: Ecology and Management of Savanna Heterogeneity*, edited by Johan T. du Toit, Kevin H. Rogers, and Harry C. Biggs (Washington, DC: Island Press, 2003): 332–48.

35. Peder Anker, *Imperial Ecology: Environmental Order in the British Empire, 1895–1945* (Cambridge: Cambridge University Press, 2001); Jane Carruthers, "Conservation and Wildlife Management in South African National Parks, 1930s–1960s," *Journal of the History of Biology* 41, no. 2 (2008): 203–36; and Jane Carruthers, "Influences on Wildlife Management and Conservation Biology in South Africa c.1900–c.1940," *South African Historical Journal* 58, no. 1 (2007): 65–90.

36. Whyte, Van Aarde, and Pimm, "Kruger's Elephant Population," 337–38.

37. Ibid., 339.

38. Piet Meiring, *Kruger Park Saga* (N.p.: n.p, 1976), 102.

39. Richard E. Leakey, *Wildlife Wars: My Fight to Save Africa's Natural Resources* (New York: St. Martin's, 2001), 220–21.

40. Jeremy Gavron, *The Last Elephant: An African Quest* (London: Flamingo, 1993), xii, 132–34.

41. Piet Meiring, *Behind the Scenes in Kruger Park* (Johannesburg: Perskor, 1982), 68–69.

42. Roderick Nash, *Wilderness and the American Mind*, 3rd ed. (New Haven, CT: Yale University Press, 1982), 9.

43. U. De V. Pienaar, "Management by Intervention: The Pragmatic/Economic Option," in *Management of Large Mammals in African Conservation Areas*, edited by N. Owen-Smith (Pretoria: HAUM, 1983), 23.

44. V. De Vos, R. C. Bengis, and H. J. Coetzee, "Population Control of Large Mammals in the Kruger National Park," in *Management of Large Mammals in African Conservation Areas*, 213–31.

45. Eleanor Momberg, "Outrage Greets Lifting of Culling Ban," *Sunday Independent*, March 2, 2008.

46. Bell's views are expressed in Adams and McShane, *The Myth of Wild Africa*, 101. Schmidtz's ethics are discussed in Lötter, "Ethical Considerations in Elephant Management," 427–28.

47. See J. J. Blanc et al., *African Elephant Status Report* (Occasional paper of the IUCN Species Survival Commission No. 33, 2007). There are more than 100,000 elephants in Zimbabwe, 130,000 in Botswana, between 14,000 and 20,000 in Mozambique, and about 20,000 in South Africa.

48. Lyall Watson, *Elephantoms: Tracking the Elephant* (Johannesburg: Penguin, 2003).

49. Rothfels, "Elephants, Ethics, and History," 117.

Bibliography

Adams, Jonathan S., and Thomas O. McShane. *The Myth of Wild Africa: Conservation without Illusion*. Berkeley: University of California Press, 1992.

Alpers, Edward A. "The Ivory Trade in Africa: An Historical Review." In *Elephant: The Animal and Its Ivory in African Culture*, edited by Doran H. Ross, 349–60. Los Angeles: University of California Press, 1992.

Anker, Peder. *Imperial Ecology: Environmental Order in the British Empire, 1895–1945*. Cambridge: Cambridge University Press, 2001.

Armstrong, Susan J., and Richard G. Botzler, eds. *The Animal Ethics Reader*. London: Routledge, 2003.

Blanc, J. J., R. F. W. Barnes, G. C. Craig, H. T. Dublin, C. R. Thouless, I. Douglas-Hamilton and J. A. Hart. *African Elephant Status Report*. Occasional paper of the IUCN Species Survival Commission No. 33, 2007.

Carruthers, E. J. "Changing Perspectives on Wildlife in Southern Africa, c.1840 to c.1914." *Society and Animals* 13, no. 3 (2005): 183–99.

———. "Conservation and Wildlife Management in South African National Parks, 1930s–1960s." *Journal of the History of Biology* 41, no. 2 (2008): 203–36.

———. *Game Protection in the Transvaal, 1846 to 1926*. Pretoria: Government Printer, 1995.

———. "Influences on Wildlife Management and Conservation Biology in South Africa c.1900–c.1940." *South African Historical Journal* 58, no. 1 (2007): 5–90.

———. *Wildlife and Warfare: The Life of James Stevenson-Hamilton*. Pietermaritzburg: University of Natal Press, 2001.

Cumming, R. Gordon. *A Hunter's Life in South Africa*. Johannesburg: Galago, n.d. (first published 1850).

Delegorgue, Adulphe. *Travels in Southern Africa.* 2 vols. Durban: Killie Campbell Africana Library, 1990, 1997.

De Vos, V., R. C. Bengis, and H. J. Coetzee. "Population Control of Large Mammals in the Kruger National Park." In *Management of Large Mammals in African Conservation Areas,* edited by R. N. Owen-Smith, 213–31. Pretoria: HAUM, 1983.

Fitzsimons, F. W. "The African Elephants." In *The Natural History of South Africa,* Vol 3., *Mammals,* 242–76. London: Longmans Green, 1920.

Garner, Robert. *Animals, Politics and Morality.* 2nd ed. Manchester, UK: Manchester University Press, 2004.

Garrard, Greg. *Ecocriticism.* London: Routledge, 2004.

Gavron, Jeremy. *The Last Elephant: An African Quest.* London: Flamingo, 1993.

Gowdy, Barbara. *The White Bone.* London: HarperCollins, 1999.

Gray, Stephen. *Southern African Literature: An Introduction.* Cape Town: David Philip, 1979.

Hall-Martin, Anthony J. "Distribution and Status of the African Elephant Loxodonta africana in South Africa, 1652–1992." *Koedoe* 35, no. 1 (1992): 65–88.

Hammond, Dorothy, and Alta Jablow. *The Africa That Never Was: Four Centuries of British Writing about Africa.* New York: Twayne, 1970.

Harris, William Cornwallis. *Portraits of the Game and Wild Animals of Southern Africa.* London: n.p., 1840.

———. *The Wild Sports of Southern Africa.* London: Henry Bohm, 1852.

Hoffman, M. Timm. "Major P. J. Pretorius and the Decimation of the Addo Elephant Herd in 1919–1920: Important Reassessments." *Koedoe* 36, no. 2 (1993): 23–44.

Jamieson, Dale. *Ethics and the Environment: An Introduction.* Cambridge: Cambridge University Press, 2008.

Kruger, Paul. *The Memoirs of Paul Kruger.* 2 vols. London: T Fisher Unwin, 1902.

Leakey, Richard E. *Wildlife Wars: My Fight to Save Africa's Natural Resources.* New York: St. Martin's, 2001.

Lötter, H. P. P. "Ethical Considerations in Elephant Management." In *Elephant Management: A Scientific Assessment for South Africa,* edited by R. J. Scholes and K. G. Mennell, 370–405. Johannesburg: Wits University Press, 2008.

Luxmoore, R. "The Ivory Trade." In *The Illustrated Encyclopedia of Elephants,* edited by S. K. Eltringham, 148–57. London: Salamander, 1991.

Lye, William F., ed. *Andrew Smith's Journal of His Expedition into the Interior of South Africa, 1834–1836.* Cape Town: Balkema, 1975.

MacKenzie, John M. *The Empire of Nature: Hunting, Conservation and British Imperialism.* Manchester, UK: Manchester University Press, 1988.

Meiring, Piet. *Behind the Scenes in Kruger Park.* Johannesburg: Perskor, 1982.

———. *Kruger Park Saga.* N.p.: n.p, 1976.

Meredith, Martin. *Elephant Destiny: Biography of an Endangered Species in Africa.* New York: Public Affairs, 2001.

Moran, Joe. *Interdisciplinarity.* London: Routledge, 2002.

Murray, John A., ed. *Wild Africa: Three Centuries of Nature Writing from Africa.* New York: Oxford University Press, 1993.

Nash, Roderick. *The Rights of Nature: A History of Environmental Ethics.* Madison: University of Wisconsin Press, 1989.

———. *Wilderness and the American Mind.* 3rd ed. New Haven, CT: Yale University Press, 1982.

Orwell, George. *Shooting an Elephant and Other Essays.* London: Secker and Warburg, 1950.

Parker, Ian. *What I Tell You Three Times Is True: Conservation, Ivory, History and Politics.* Moray, UK: Librario, 2004.

Parker, Ian, and Mohammed Amin. *Ivory Crisis.* London: Chatto and Windus, 1983.

Payne, Katie. *Silent Thunder: Hidden Voice of Elephants.* London: Weidenfeld and Nicolson, 1998.

Pienaar, U. de V. "Management by Intervention: The Pragmatic/Economic Option." In *Management of Large Mammals in African Conservation Areas,* edited by R. N. Owen-Smith, 23–26. Pretoria: HAUM, 1983.

Pratt, Mary Louise. *Imperial Eyes: Travel Writing and Transculturation.* London: Routledge, 1992.

Pretorius, P. J. *Jungle Man: The Autobiography of Major P. J. Pretorius C.M.G. D.S.O. and Bar.* Sydney: Australasian Publishing, 1948.

Ritvo, Harriet. *The Animal Estate: The English and Other Creatures of the Victorian Age.* Cambridge: Harvard University Press, 1987.

———. "History and Animal Studies," *Society and Animals* 10, no. 4 (2002): 403–6.

———. *The Platypus and the Mermaid and Other Figments of the Classifying Imagination.* Cambridge: Harvard University Press, 1997.

Roche, C. J. "'The Elephants at Knysna' and 'The Knysna Elephants': From Exploitation to Conservation; Man and Elephants at Knysna, 1856–1920." B.A. (Hons) dissertation, University of Cape Town, 1996.

Rothfels, Nigel. "Elephants, Ethics, and History." In *Elephants and Ethics: Towards a Morality of Coexistence,* edited by Christen Wemmer and

Catherine A. Christen, 101–19. Baltimore: Johns Hopkins University Press, 2008.

Scholes, R. J., and K. G. Mennell, eds. *Elephant Management: A Scientific Assessment for South Africa.* Johannesburg: Wits University Press, 2008.

Selous, Frederick Courteney. *A Hunter's Wanderings in Africa.* Johannesburg: Galago, 1999 (first published 1881).

Skead, C. J. *Historical Incidence of the Larger Land Mammals in the Broader Eastern Cape.* 2nd ed. Edited by A. F. Boshoff, G. I. H. Kerley, and P. H. Lloyd. Port Elizabeth: Centre for African Conservation Ecology, Nelson Mandela Metropolitan University, 2007.

Thomas, Keith. *Man and the Natural World: Changing Attitudes in England, 1500–1800.* Harmondsworth, UK: Penguin, 1984.

Thorbahn, Peter F. "The Precolonial Ivory Trade of East Africa: Reconstruction of a Human-Elephant Ecosystem." PhD dissertation, University of Massachusetts, 1979.

Watson, Lyall. *Elephantoms: Tracking the Elephant.* Johannesburg: Penguin, 2003.

Wemmer, Christen, and Catherine A. Christen, eds. *Elephants and Ethics: Towards a Morality of Coexistence.* Baltimore: Johns Hopkins University Press, 2008.

Whyte, Ian J. "Conservation Management of the Kruger National Park Elephant Population." PhD dissertation, University of Pretoria, 2001.

Whyte, Ian J., Rudi van Aarde, and Stuart L. Pimm. "Kruger's Elephant Population: Its Size and Consequences for Ecosystem Heterogeneity." In *The Kruger Experience: Ecology and Management of Savanna Heterogeneity,* edited by Johan T. Du Toit, Kevin H. Rogers, and Harry C. Biggs, 332–48. Washington, DC: Island Press, 2003.

Woodward, Wendy. *The Animal Gaze: Animal Subjectivities in Southern African Narratives.* Johannesburg: Wits University Press, 2008.

Wylie, Dan. *Elephant.* London: Reaktion, 2008.

Chapter 4

Keeping the Rhythm, Encouraging Dialogue, and Renegotiating Environmental Truths

Writing in the Oral Tradition of a Maasai Enkiguena

Mara Goldman

> Accounts of a "real" world do not . . . depend on a logic of "discovery," but on a power-charged social relation of "conversation." The world neither speaks itself nor disappears in favour of a master decoder. The codes of the word are not still, waiting only to be read. . . . [T]he world encountered in knowledge projects is an active entity.
>
> **Donna Haraway,** *Simians, Cyborgs, and Women*

> *Metumo ildoinye, kakai tumo iltunganak*
>
> Maasai proverb: "Mountains don't meet, but people do"

UNDERSTANDING CHANGE and continuity in African environments has always involved storytelling. As scholars of African history, anthropology, and geography have well illustrated, history has shown a disproportionate privileging of a particular kind of story: the heroic tales of Europeans and Americans exploring, studying, managing, and otherwise "knowing" Africa.

The narratives inspired by such adventure-cum-science kept African people and African voices out of the story despite the reliance of many of the authors on African informants, African laborers, and sometimes African writers, intellectuals, and politicians for the stories they were transcribing for a global audience.[1] As African historians have argued, the methods and epistemologies of such "scientific" narratives were "largely divorced from any notion that Africans could themselves comprehend, interpret, and narrate their worlds in a usable manner."[2] Indeed, the "'science' of such studies was substantially constructed around the notion of the absence of indigenous intellectual and scientific authority" (White, Miescher, and Cohen, *African Words*, 4–5). And I would add that such stories were also proposed as finished products—truth statements about African environments and societies.

As the challenges to such colonial narratives has grown over the last century, so have the questions regarding the types of narratives that should be constructed about African environments: by whom, for whom, with what methods, and in what formats? These questions are not new, having been raised by historians, anthropologists, and African intellectuals and authors for quite some time. As Ngugi eloquently argued in 1981 in *Decolonizing the Mind*, the very language that African literature is written in is important and has been debated by African writers with earnest since the beginning of the end of the colonial era. Ngugi's own choice to write only in his native tongue was a strong statement about the importance of language and presentation in expressing African ideas, knowledge, and stories, that is, in talking about African environments, both natural and social. Even translated into English, Ngugi's writing overflows with and very much depends on African (Kikuyu/Swahili) words, proverbs, and storytelling techniques. These techniques not only give voice to his ideas and knowledge but also provide a rhythm to the story.

In this chapter, however, I am less interested in African written literature than I am in African oral expressions, expressions that may or may not be recognized as literature but can play a powerful role in shaping knowledge production and power relations. I am interested specifically in how we (as scholars interested in human-environment relations in Africa) can draw from such traditions or forms of knowledge expression and negotiation to talk about, debate, and present knowledge on African environments in a way that reflects the very multiplicity and complexity inherent in such knowledge. This focus reflects my own desire to find both an indigenous format to reflect indigenous knowledge constructions in the places where I work and a style to reflect multiplicity and complexity in knowledge about natural-social systems in the written work I produce. As such, this chapter reflects several important shifts within the social sciences to embrace the multiplicity of knowledge, the complexity of natural-social systems, and the role of performance in shaping not only knowledge expressions but also power relations and knowledge constructions.[3] The increased interest in the performance, meaning, and complex relations tied to spoken statements (between different speakers and the audience, for instance) reflects a move toward a more nuanced and sophisticated understanding of oral literature and its role in society. A focus on "the relation between performer and audience in the moment of the performance" highlights the "interactive dimension between texts" (i.e., the written form of spoken statements; Furniss and Gunner, *Power, Marginality,* 2) rather than seeing the text or narrative as truth in and of itself.

Much of the work in political ecology on African environments has focused on the power of narratives that either ignore African voices and knowledge of their environments or that paint them in an entirely negative light. Some have suggested the need to produce counternarratives to create better, more powerful stories about African environments.[4] But what sort of counternarratives should we tell? And how will the voices, knowledge, and actions of the different African individuals and communities we work with in different environments come through in the new narratives that are created? I suggest that the answer lies in moving beyond Western ideas of narrative itself. In many ways a narrative implies storytelling, a fuzzy space between fact and fiction (Haraway, *Simians, Cyborgs*). Yet narratives can also become solid, lasting, and seemingly true in and of themselves, as we have seen with the colonial-inspired narratives of African "bush" environments and pastoral peoples in particular. Narratives, when written, often become a form of truth. I am interested in taking the counternarrative idea further: opening up scientific truths to narrative writing style and subjecting narrative to the fluidity it sees in oral forms. I am, as the title of this chapter suggests, interested in creating rhythm in the written form, a rhythm of words, ideas, and truth claims.

As my own work takes place in Maasai areas in Tanzania and Kenya, it is the rhythm of Maasai communication that I draw from. I am interested in building from Maasai oral traditions and communication techniques to talk about land use and environmental management in Maasai areas. Yet I am not interested in just telling new stories perhaps from a Maasai "indigenous" perspective or that reflect a counternarrative to that promoted through explorer, colonial, and scientific literature. I am instead interested in working with a different medium, a medium of dialogue, debate, and negotiation. For while Maasai do tell stories—oral narratives rooted in the past and modified by the narrator differently for different contexts—they spend a lot of their time negotiating, discussing, and debating in meetings (the *enkiguena;* pl.

inkiguenat). Living in Maasai villages and trying to get research done, I too have spent a lot of time at meetings, where I watched all sorts of issues being discussed, negotiated, debated, and decided upon. The *enkiguena*, I learned, maintains a certain sense of rhythm, reflects a particular structure, and upholds certain ideals of communication. I suggest that the *enkiguena* structure provides a model framework for discussing multiple knowledge constructs about the environment, those from members of Maasai society as well as those from the scientific domain of storytelling. This chapter is a description of why I think this is the case and a presentation of my own attempts to write in the Maasai oral tradition of the *enkiguena*.

This chapter is divided into three parts. The *enkiguena* is presented as a formal structure of Maasai communication, reflecting patterns and rhythm as seen in other less formal Maasai verbal exchanges (eating the news).[5] I argue that the *enkiguena* ideals provide a template for promoting dialogue as well as a theoretical construct and presentation format to talk about multiplicity and complexity in African natural-social systems. The remainder of the chapter illustrates how I utilize the *enkiguena* structure to talk about wildlife-livestock interactions in a particular place. With this example I hope to illustrate the way that an *enkiguena* works in practice and how the structure—its rhythm, format, and ideals—can be utilized as a format for presenting complex, multiple, and often incommensurable knowledge constructions between, across, among, and within Maasai and scientific knowledge domains.

My goal in the final format—which was a dissertation and will soon become a book—is to reduce my own role in the story. I am the compiler but not the storyteller. I glue pieces together but let the voices come through on their own. My success at doing this was recently noted when I was told by someone who wanted to cite the dissertation that she did not know how to cite it. Should she cite me, the author of the final product, or should she cite the person who was "speaking" in the *enkiguena* whose ideas she was using? It was not clear. This was a sign for me that my writing style had succeeded at blurring the lines of expertise and authority that we are so accustomed to positioning with the scholar rather than with our informants and partners in research.

Keeping the Rhythm:
Maasai Communication and the Art of Speech

There has been much debate about the role of oral literature in African history and in other parts of the world where much of the knowledge about the local environment is expressed and remembered orally. Anthropologists in particular have stressed not only the need to understand *what* is said—in the sense of finding new sources of data—but also the need to pay attention to *how* things are said. As Julie Cruikshank has put it (or rather as her informants put it to her): it is about "getting the words right."[6] She is interested in how the words inscribed in place-names are told and remembered. She helped the people she worked with to write down the words so they would not be lost. Yet other anthropologists have warned of the difficulties and dangers of writing up oral knowledge. Spoken words are transient; writing, they warn, endures. Spoken words are active and shared, whereas writing is static and somehow "owned."[7] "Writing things up," they warn, "gives authority to a particular view and a particular writer."[8] My response to these completely valid and important concerns is that *not all writing is the same.*[9] This can be seen in some of the writings of Cruikshank herself, who often presents full texts of stories or interview transcripts, and can also be seen in the playful writing of Science and Technology Studies (STS) scholars.[10] For myself, writing in a way that mimics the Maasai oral tradition of the *enkiguena,* I attempt to re-create the web and flow of dialogue that occurs in an active *enkiguena* to present multiple knowledge claims. In this way I am interested in trying to do more than "get the words right"; I also want to appreciate the "art of speech" as expressed in words, in speech patterns, and in the rhythm of dialogue.

The *enkiguena* is the most formal setting where the art of speech is practiced by Maasai. A similar pattern of speech is visible in a more regular daily ritual, that of eating the news. In Maasailand, to share the news with someone, to sit and catch up with an old friend, or to find out more about a visitor is to eat the news (*ainos ilomon*). News is "eaten" because it is not just passed back and forth but instead is presented by one person and then taken in, digested, and ruminated on by another.

To eat the news is indeed a ritual of sorts with a structure and a rhythm. The process is initiated when one of the people involved asks one of a series of possible questions regarding the other's condition or state of being, such as "You are well?," "You are not sick?," or "You are rested?" This signals the other person to begin talking, and so he or she does. She/he answers the question always in the positive, regardless of his/her state of being: "Yes we are well," "No I am not sick," "We are well rested." She/he then repeats this notion in multiple ways: "We are well/good (*kirasidaan*)," "We are at peace (*kiserian*)," "We are all healthy (*kirabioto*)," "There is nothing wrong (*miti entokin tohrono*)," and so on. After this refrain, she/he really begins to talk: "the children all have malaria, the cattle are all sick, the dry season has hit hard," explaining, of course, that all is *not* well. Nonetheless, and regardless of how bad a person's condition might in fact be, she/he continues to repeat throughout the conversation, as a chorus of sorts, the various ways in which she/he is fine.[11] The more talented an orator the speaker is, the longer and more musical and rhythmic is the news. The rhythmic chorus of some variation of "we are at peace" and "we are all well" is divided by long periods of detailed talk about the topic of interest: the real state of affairs.

The entire time that the speaker is talking, the other participant in the dialogue is just listening. This is not a back-and-forth conversation. The one listening is not to speak except to acknowledge that she/he is listening by "responding" accordingly, with the affirmative sounds of *ooó* or *eeé* or with a quick word, *tedo?* ("you say so?") or *esipa* ("it is true"). Failure to do so suggests that one is not listening. If the listener does to respond in this way, she/he is reminded to—*tedo ooó* ("say *ooó*") or *tedo eeé* ("say *eeé*")—by the speaker. This responsive listening is so central to Maasai communication that young children who are learning to talk can be heard telling one another "*tedo ooó*" or "*tedo eeé.*" When the speaker is finished sharing her/his news, she/he makes it clear by calling out a particular long tone, raised at the end. Then and only then does the listener know that the speaker is finished and that it is now his/her turn to speak. And so the speaker and the listener now switch roles. There is no need to ask the opening question "You are well?" again; the listener has been cued that it is now his/her turn to speak, to share his/her news, and does in the same manner as described above.

This formal sort of news exchange can occur between men and women of different ages and social status, all with the right to their own time to speak. There are of course complex rules of social interaction among Maasai that cross gender and age lines and are tied to kinship and respect (*en-kanyit*) so that certain individuals (men and women) are socially restricted from initiating the news with others. Maasai will often equate respect directly with fear: fear of initiating certain interactions, fear of showing disrespect. Despite these complex social relations, all Maasai adults eat the news with each other, even if they are afraid because of respect to initiate the process. Thus, in theory and in practice they all have a turn to speak.

The *Enkiguena*

The rhythmic pattern of speech, mutual respect, and shared time for speaking and listening that occurs when Maasai eat the news occurs, at a much more formal level, at an *enkiguena*. The *enkiguena* illustrates the strength of the spoken word and the importance of debate (for making decisions and maintaining social order) within Maasai society. The *enkiguena* is central to Maasai governing organization and leadership structures and is the tool of arbitration used to settle conflicts, make decisions, and organize events.

The centrality of the *enkiguena* is illustrated by the fact that *olaigwenani* (pl. *llaiguenak*), the word for "leader," shares the same root as the word *enkiguena*. The literal translation of *olaigwenani* is "spokesman," although *llaiguenak* are usually referred to as "leaders."[12] Leadership qualities (such as knowledge and respect) are intimately associated with communication skills. The customary role of the elders as the decision makers and community leaders, combined with their love of the spoken word and skills at arbitration, means that most *enkiguena* are called by and/or dominated by the elders. This would explain why the term has been often (mis)translated as a "council of the elders." An *enkiguena*, however, is just a meeting, and an *enkiguena* can be called by young men (*ilmurran;* sing. *olmurrani*),[13] women, clan members, and village government as well as by elders.

The *Enkiguena* Ideal: Context, Format, and Principles

There are three primary principles underlying the *enkiguena* structure that make it an ideal structure for expressing multiplicity. First, discussion is opened by an agenda that names *only* the problem at hand (e.g., the *issue* "concerning" the dam, the *issue* "concerning" the school).[14] There is no mention of possible solutions within the problem statement itself or by the person who opens the meeting (e.g., to vote if we should place the dam at location X, to decide if Y should be fined X number of cattle). Second, *everyone* at an *enkiguena* has the freedom (theoretically) to speak and to be listened to; there are no rules constraining a person's contribution. And finally, a meeting is finished only when a consensus has been reached.

At an *enkiguena*, each speaker speaks until she/he is finished regardless of how long this might take; then and only then does another person stand to speak. The one who speaks does not need to respond to or in any way address what the last speaker said (although she/he often will); it is just the current speaker's turn to explain her/his story, just like when eating the news. And for the most part, everyone's experience or knowledge is respected and considered relevant, which means that a meeting can take all day, much gets repeated, and the conversation may go in unexpected directions.[15] This process was explained by a Maasai junior elder to his *mzungu* (Euro-American; pl. *wazungu*) guest who was clearly confused trying to follow a meeting that was being conducted in Maasai style:

> We do things the Maasai way, we do not know how to make things short [*kufupisha*], like you *wazungu*. We can take a very long time, but it is like this: one person goes into detail

[*anachuja*], another makes an error [*anakosea*], another goes far off the topic [*anaenda mbali*], and then we return right where we started. If you try to have 10 agenda items in one meeting, [it is not possible]. It is not like those [meetings] of *wazungu,* or even ... those others who have developed [*watu walioendelea*]. You can have only two agenda items and still it will take a very long time. Until most of the people present have stood.

An individual can stand to speak multiple times, with something new to say each time, as the dialogue progresses, twists, and turns and a consensus is constructed.

So how is a consensus reached? There are no choices put forth, and there is no voting. The dialogue rarely follows a linear path but instead circles around and meanders off in different directions as different speakers contribute new ideas, knowledge, and opinions to the topic at hand. People will often draw on memory and discuss how a similar problem was tackled in the past, what they decided on then, and if that is relevant today. The consensus is not reached magically; there are inner workings regarding key players, skilled arbitrators, and decisions about who to trust and who to respect. Importantly, a consensus is *built* (enacted), not voted on (as in discovered). As such, the consensus encapsulates the multiplicity of views, knowledge, and opinions voiced by those present and is final when no one has anything left to say. In other words, a consensus does not represent a singular (cohesive, complete) Maasai viewpoint but instead is the piecing together of multiple contributions from partially connected participants.[16] The participants are partially connected in that they are all at once Maasai, but they are women and men of different age sets, wealth, clans, and status within Maasai society reflecting different social loyalties, kinds of knowledge, expertise, and experience and commanding different degrees of respect. I therefore suggest that an *enkiguena* can also be used to piece together different knowledge contributions of Maasai and non-Maasai and to challenge perceived truths of the environment and African land use.

Enkiguena in Theory: Capturing Complexity and Maintaining Multiplicity

> The discovery of multiplicity suggests that we are no longer living in the modern world, located within a single *epistème*. Instead, we discover that we are living in different worlds. These are not worlds—that great trope of modernity—that belong on the one hand to the past and on

> the other to the present. Instead, we discover that we are living in two
> or more neighboring worlds, worlds that overlap and coexist.
>
> John Law and Annamarie Mol,
> *Complexities: Social Studies of Knowledge Practices*

At an *enkiguena*, truths are negotiated. This is done in such a way that a consensus is built that is acceptable to all, while multiplicity (difference) is acknowledged, encouraged, exposed, and recalled. The consensus as such is not a finely crafted singularity that denies the existence of the various voices and knowledge claims that led to its creation. The consensus records multiplicity through the dialogue in which it is enacted and as such keeps an open and accountable history of how the consensus was reached (i.e., the traces of the network and identities of the allies are exposed).[17] For instance, Maasai will often discuss the outcome (consensus) of a meeting together with the various contributions made, revealing the messiness (politics, social relations, history) behind the final consensus. The messiness, localness, and contingency involved in the *negotiation* of truth, is not erased, as is often the case in scientific representation.[18] As STS scholars have well argued, all knowledge constructions are local and site-specific, yet "scientific knowledge is publicly presented as universal and rational."[19] So, according to Turnbull (*Masons, Tricksters*, 13, my emphasis),

> given the lack of universal criteria of rationality, the problem of working disparate knowledge systems together is one of *creating a shared knowledge space* in which equivalences and connections between differing rationalities can be constructed. *Communication, understanding, equality* and *diversity* will not be achieved by others adopting Western information, knowledge, science and rationality. It will only come from finding ways to work together in joint rationalities and in knowledge spaces constituted through these joint rationalities.

I suggest that a focus on the ideal structures and principles underlying an *enkiguena* can enable the creation of such a space in three important ways: by maintaining tension between sameness (unity/consensus) and difference (multiplicity), by following the general laws of symmetry[20] within and across different knowledge spaces (e.g., the words of all), and by expanding the realm of accepted methods (voices, views) into a method (knowledge) assemblage. This final point refers to the acceptance of multiple methods or ways of knowing within one dialogue. In this sense the *enkiguena* represents a nonlinear framework, an open-ended agenda, and an acceptance of all voices. I propose the *enkiguena* as an African format for renegotiating environmental truths, one that is true to the rhythm of oral dialogue and debate.

By using an *enkiguena* structure to engage the knowledge contributions of scientists and Maasai in particular contexts (e.g., in discussing wildlife use of village lands), we can expose (and create) the *overlap* and *interference* between these different knowledge spaces, in theory and practice. The *enkiguena* format keeps us from having to choose between difference and sameness, multiplicity and singularity, within and between Maasai and scientific contributions.

In using an *enkiguena* format to *write* about my research findings, I am also making a political point by choosing the context, metaphor, and rhetoric of those usually being Othered.[21] Choosing a format of oral decision making and storytelling to present scientific data reverses the historical pattern of using Western formats to suppress or (mis)represent African knowledge. And as Harding (*Is Science Multicultural*, 222) aptly questions, "after all, why should other cultures' projects have to be named in European terms in order to be taken seriously by Europeans?" Why not use a local (Maasai) practice, the *enkiguena,* as a method assemblage—expanded from its customary usage among Maasai to capture diverse knowledge claims from Maasai and conservation scientists across differences of method, practice, gender, age and status—to talk about wildlife in Maasai lands?

I now present an example of writing my findings as an *enkiguena*. I did this by raising issues such as where the wildlife are, wildlife-livestock relations, and conservation corridors as agenda items in an *enkiguena* that included Maasai participants, my own scientific data collection, and the data and opinions of scientists, managers, and conservationists. Staying accountable through citations and footnotes, I wrote playfully, using stage directions, descriptions, and creative acts of speech to maintain a sense of performativity, that is, to keep the rhythm and present my data, as well as the data and thoughts of others, through dialogue. When possible and appropriate I pulled from actual meeting dialogues and settings and then added interview data and brought to life the written words of past researchers. Pseudonyms are used for all interview data, but the age set is provided in the name, reflecting the place of the individual within society (as a warrior [*olmurrani*], junior elder [*landis*], or elder for men and as young married woman, mother [*yeiyo*], or grandmother [*koko*] for women).[22] Social norms suggest that women, the young, and the poor do not have the courage to speak in front of elder men and that wisdom and respect is thought to follow with age and reputation, although that is not always the case.

Enkiguena Agenda: Competition, Cooperation, or Coexistence?

The Setting

Today's *enkiguena* is in the dry riverbed (*korongo*), under the shade of an old ficus tree. The preferred location is under the large baobab (*olmesera*) tree that stands at the administrative center of the village. However, it is now the end of a long dry season, with most trees bare and shade hard to come by. The large ficus tree in the *korongo* provides the rare escape from the hot October sun. Some men lie outstretched leisurely on the sloping walls of the *korongo*, while others are crouched on the roots of the trees or squatting on rocks on the *korongo* floor. Women gather as a group in the tall grass at the top of the *korongo* out of direct sight line of the men. Invited guests (scientists, local conservation managers) either sit among the men or lean against their cars, parked where the road passes through the *korongo* floor.

Note: I start with a transcript of a meeting in which a contract with a tour company was discussed (i.e., the need to agree to the contract and the dam that the company was going to build). I use this as a starting point to talk about wildlife-livestock grazing impacts on the range, a discussion that began at this meeting. I skip (from the original text) the lengthy start of the meeting as well as discussions of how much should be repeated from the last meeting and attempts by Chris, a businessman/conservationist and the owner of the tour company, to get his papers signed so that he can go. Just as it seems that this will happen and the meeting will end shortly, Landis T stands to speak.[23]

LANDIS T: [*A healthy and strong man, he is well respected and trusted by most villagers. He is not around much these days as he works as a game scout in the new neighboring conservation area, the Manyara Ranch (MR).[24] He is very smart and very diplomatic. . . . He speaks loudly to be heard over the noise.*] "I would like to say before we finish the meeting, I have an issue to raise that I think we need to discuss. What I would like to talk about is that this dam that is to be built, won't it also be used by wildlife? You all know as I do that if there is water here all the wildlife will come [into the village]. They will all come from the MR. And what will happen when all of the wildlife fill this space? What will we talk about then? I am very thankful to our guests for the help they bring, it is very good to have this dam, but we need to talk with the MR so that they know that the wildlife will all come to our village land and if they come to graze, when the grass finishes what will the cattle eat? Will we be able to negotiate use of the MR for grazing? These are all very important questions that we should discuss before we agree officially to anything."[25] [*Others stand up and talk about this . . . men of various ages and affiliations. . . . Others (village government members and Chris) are clearly getting angry.*]

MAKAA SI: [*An easygoing and well-liked man, he is often at odds with the village government, as he is not afraid to say what he thinks. He is poor and has no position of authority but is trusted and respected for his intelligence and honesty.*] "What Landis T says is very true. In a short while in the future there will be nothing (no pasture) left for the cattle. The area will be full of wildlife, and they will finish the grass and leave the ground bare. This is not a light matter, and we have not had enough discussion about this. We need to discuss this more. We discussed these issues when we were looking for the best place to build the dam. Not in a grazing area, we decided, as the grass will be finished quickly by cattle and wildlife. Especially wildebeest; they are particularly damaging [*maharabifu sana*]. They will finish all the grass." [*Sounds of "oh-hoo" and "ooshoo" . . . from the men and women around the circle, attesting to how the wildebeest will "finish the grass."*]

LANDIS Z:[26] "This is true that we need to look carefully before we build a dam. What this elder says is true. If a dam is built, and there is water year round? . . . *Oshooo*, all the animals in the MR will come to the village lands. They will also come near to the *bomas* [settlements].[27] Only elephants will stay in the ranch because of the trees there."

CHRIS: [*Chris is supportive and says that he has already started to talk to the MR about this issue, but he pushes for the paperwork to be prepared so that he can leave.*]

SENIOR ELDER: [*Despite the shift in subject and attempts to close up the meeting so that Chris can go, a senior elder stands to speak, holding tightly onto a tall thin walking stick.*] "Really it is true. If all these dams are built— this one Chris is planning to build us, the one in Esilalei [Ranch dam], which is being fixed—then this whole area will just be a conservation area [*hifadhi tu*]. And there, there and, there [*he points with his free hand in the direction of the ranch, TNP, and LMNP*] it is conservation [*hifadhi*] already!"[28]

[*There is commotion from Chris and members of the village government. "What are these people doing?" they mutter; everything was already decided on.*]

LANDIS M: [*A natural arbitrator, he tries to reconcile the views or at least calm down Chris and his allies. The chairman quiets down the crowd so that*

Landis M can speak.] "Nobody disagrees [to the campsite or the dam], but we need to have a good system for planning. The village will become full of wildlife, yes. But isn't this better than not having water? This is a big question, it is a big issue."[29]

[*End of transcript from meeting.*]

YEIYO N: [*The women are talking loudly among themselves, and they are called on by the chairman to speak. . . . One does so quickly and quietly.*] "We agree the dam should be built where proposed. There are places that we know the elders were discussing for the dam which are no good, because the water there is salty [*magadi*]."[30] [*She trails off and is quiet while other women whisper to her. An elder begins to stand up to speak, and just then Yeiyo N continues to speak, loudly this time.*] All this talk about wildlife is fine. But like Landis M said, even if the village is full of wildlife won't it be better to have water? We say *yes* [*says this last word with extra emphasis, and the other women laugh*]. We struggle to find water here every day to drink and cook with. Don't give up this dam!"[31]

KOKO B: [*As the elders try to speak, another woman, weathered with age, speaks up in a loud, coarse voice.*] It is true. "When the dam is built, all the animals will come. All of them [*using her hands she gestures in a coming motion*]; the whole area will be *full* of wildlife, they will all come. Oh, and these animals will be very happy because this was their area. It is open plains [*mbuga wazi*]; they like it here. They will be happy because they will get water and they won't be thirsty again. And people too. All those Maasai that moved to look for water? They will come back, *together* with the wildebeest!" [*The women all laugh.*][32]

PETER:[33] [*The microenterprise development specialist for the African Wildlife Foundation (AWF), which runs the MR. He enters the center of the meeting, from where he had been leaning on his car, and turns and looks at everyone before speaking in a very authoritative tone.*] Do you want the dam or not? It sounds like you want it but then only if you can get something in return. "[You just] want access to the ranch and now [you] see an opportunity to pull [your] arguments . . . so as to gain access. The Manyara Ranch is an island. The rest of the area is all overgrazed, badly overgrazed. [You] need to use [the land] within its capacity and not have more animals than the land can take." So is it possible for you to access the ranch if wildlife use of village lands increases? I'd have to say "yes and no. Yes, if [you] learn to keep livestock in a sustainable way. What [you] do now is not sustainable; it is not a good match of livestock numbers with the resource base [i.e., carrying capacity]. So, no [you should not be able to graze in the ranch], or five years from now the ranch will look like it does here, where there is no grass." [*He looks around and swings his arm at the nearly bare short grass plains behind him. There is a great deal of commotion in the crowd until a highly respected elder stands to speak.*]

SEURI K: [*Chuckling to himself, he puts on a serious face as he addresses his words at Mr. Peter in particular.*] First of all, this area that you point to as being overgrazed, that is what we call *olpura alamunyani*! I have lived in this area all my life, and it has always been like that, "even before they drew the boundary for the ranch, it was like this." You see *olpura* means a place that is bare and no grass grows, and *alamunyani* refers to a different kind of soil. "[*Alamunyani* is] a place that above it has grass, but if you dig a foot below, you will get *salt*, which causes all these changes in the grass. Because it is not just there that is *alamunyani*, but throughout the village there are places we call *alamunyani*. Grass grows in some of these places, but it is a short, light grass that finishes fast [e.g., annuals]; in other places only *embenik* [dicots/herbs] grow, which also finish fast," which is why the area is bare now; it is the end of a long dry season. [*He pauses, looking around at the dusty landscape.*] "It is complex; there are different types of *alamunyani*." In this area you point to, grass doesn't grow, because the salt is very close to the surface. "Even long ago it was like this. [*He shakes his head.*] No grass, just dicots/herbs. Of course there *are* also impacts from grazing. If cattle are many in one area, they can finish that grass until it changes and herbs come to dominate. Even trees will increase because you can't burn since there is no grass to burn, so the trees will continue to grow. But there is no place in these villages where I have seen this happen."[34]

GAME SCOUT 1: I am sure what the elder has said is true; he knows this land better than I do. But I understand what Peter is saying, and I know that it is something that is hard for Maasai to see. "It is not that the cattle from the villages will finish the grass in the MR, but they can ruin the environment because they will come in on one path—more than 11,000 cattle from Esilalei, they will bring a *Korongo* [create gully erosion]. And then no grass will grow again. I have seen this and learned this in Mwanza, where I studied wildlife conservation.[35] Also, if cattle eat all in one place, they can finish the grass until it is taken by the wind and the land becomes tired [wind erosion]. Maasai don't see this. And they don't take into account the difference between now and long ago. For example, in 1989 there were only 150 cattle at our *boma*. Today, there are close to 3,000. If we want to conserve the environment and wildlife, we need to get rid of some cattle."[36] [*People start to make noise, upset with what he is saying. An elder from his boma, the owner of the 3,000 cattle, tells him to sit down and stands up to speak.*]

MESHUKI OM: [*He is one of the most respected and powerful elders in the area and by far the wealthiest. Despite the respect that he knows he commands, he speaks in a gentle, melodic tone*] Youth these days, they forget what their elders taught them and brag about what they learn in school. It is good for them to study to learn how to gain from the milk of wildlife like we do from cattle. But, if the young men spent more time at home and less at school, they would know that "cattle can't ruin the environment by finishing the grass, [not] if it rains." But, he was correct in talking about the damaging effects of cattle pathways, where grass cannot grow, because cattle walk there every day, "but next to [the path] grass will grow as usual. But we have rules about where the cattle can go at different times to prevent this, and we have pasture reserves [*alalili*] and even rules regarding paths. And we can tell our sons, 'Do not take the cattle to that place again today; you will finish the grass.'" So [*he chuckles*], yes, sometimes I guess cattle can finish the grass, and then sometimes when it rains, *herbs* will grow instead of grass and then the cows don't get enough to eat. "But an area can fluctuate between grass and *herbs* depending on the rains. It is really rainfall that determines these things. But it is not just cattle that can finish grass. Even wildlife can finish grass. They are all animals. Wildebeest really ruin an area because wildebeest don't leave an area until they finish the grass completely, leaving only dust behind."[37]

MAKAA MN:[38] The elder has spoken truthfully. What I say is that "there are two things that can ruin pasture. Cattle can go to one place every day, where they mix up/trample [*kuvuruga*] the grass. It is not just grazing; it is that they step on the grass every day, and the grass gets tired. Second, wildebeest can come to one place and stay, only one week, and they can finish the grass. They eat it until the bottom. Only wildebeest, no other animal does this." [*In his joking way he nods his head in imitation of a wildebeest feeding, and then, laughing, he swings up his stick to motion that he is done. He starts to walk away and then stops, addressing the crowd.*] "And farms. Farms can ruin the land. They make the land tired."

[*The discussion continues, with different viewpoints on the impact of wildlife and livestock on the range. Katherine Homewood appears, and here I draw from her work in Ngorongoro, where they found evidence of the impacts of wildebeest on the pasture but none from cattle directly, and put that in quotes. I take artistic liberty to make her words spoken and in reaction to the discussion, for instance, reacting to Mr. Peter's words.*]

KATHERINE HOMEWOOD: Such statements were common where we worked in Ngorongoro, where "[a]ccusations of overgrazing have typically been poorly defined, unsubstantiated, and based on spot judgments."[39]

[*I then bring in Dr. Raphael Mwalyosi, a professor at the University of Dar es Salaam, who is able to speak more from a local perspective, having written a lot on this area, and again I bring to life his written words.*]

RAPHAEL MWALYOSI: Even here, in the Lake Manyara area, we can see that both wildlife and livestock can impact vegetation composition as well as soil structure. For instance, the most likely cause of the decline in *Themeda trianda* and *Hyparrhenia* sp. grasses so favored by wildlife is "overgrazing by game animals and restriction of burning."[40]

[*Different Maasai speak about burning: its importance and historical use and its value in pasture management.*]

Note: There is a shift now in the dialogue away from the specific impacts of wildlife or livestock on the range toward their coexistence in the pasture. Do they avoid each other, help each other, or compete? Do they separate out by time and space, and how do predators and herd boys affect things?

GAME SCOUT 2: [*He is young and is not from this village and also studied wildlife conservation at Mwanza. He is shy but speaks clearly.*] "It is true that there is a lot of grass in the ranch. I don't think that is the reason the management doesn't want too many cattle inside. They are not just coming into the ranch for grass but also for water. But they will eat all the grass near the water, and it will become a degraded plain. Also, if people dig [wells] in the dry riverbed, when it rains the soil is easily carried away. And the animals can fall into the wells. And the elephants might use the wells but will then cover them up after using them. That is certain. The elephants always cover up the watering hole after drinking. Another thing, when cattle come into the ranch they bother wildlife. The problem is with the herd boys, who come with the cattle; they chase away wildlife, and they usually have dogs, which makes things worse."[41]

OLMURRANI LEKEI: [*Only a warrior, he is not afraid to speak after the game scout. The* olmurrani *saunters into the center of the circle and, leaning on his walking stick, speaks.*] Maybe what the game scout says is true *sometimes*, but [other times] we share our wells with wildlife. I was just out there digging wells for our goats. "And wildlife come to use the water as well, even elephants. Elephants come and use the wells we dig. Sometimes when we are still there. If they come we just move out of the way, and let them drink first and then we return to give water to our animals. There is no problem. They don't ruin the well. They do try to cover it up, like the game scout said, but this is not a big problem because we fix up the wells every day anyway. So we have no problems sharing our water with wildlife."[42] But, the ranch management does not want us there. I guess they think we are interfering with wildlife. I don't know. . . . "The other day the manager caught us there. [*He looks over at the manager with a combination of shy respect and anger before continuing.*] He said we weren't supposed to be there. He threatened to cover our wells up . . . but the *mzee* [elder of the *boma*] begged permission and now we are allowed to go there [for water]. We have no place else to go. There is no water anywhere."[43]

ELDER:[44] "Wildlife and livestock can stay together, drink the same water and eat the same pastures, as long as the wildlife have already become familiar with people. Or if they don't see people, wildlife will stay together with livestock because they don't bother each other; they are not afraid of each other. They are afraid of people. They have become afraid of people because they are chased away by herd boys, every day." And if we tell the herd boys not to chase wildlife, if for instance we were benefiting from having wildlife around, *basi* [that's it], they would stop![45]

LANDIS K: [*Sitting with MG*[46] *and M, Landis K is encouraged to stand and speak. He is usually shy around elders and is not one to speak at meetings. Since he will be speaking on the transect work, he feels confident but still speaks quietly.*] It is true that wildlife can be afraid of people, and maybe that is why during our transect work we sometimes found more animals in the Manyara Ranch than in the village. But the funny thing is that while wildlife might be afraid of herders during the day, they seek out *bomas* at night for protection against the real danger: *olowaru* [predators, usually lions]. Now it is the dry season, so all the animals are back in the park. But as soon as it rains, *ooohoooh*. . . . There are so many wildebeest by my *boma* that it's dangerous to go out at night because of all the lions![47]

MG: [*Looking through her papers and comparing tables with M, she slowly stands to speak with the papers in hand.*] What Landis K says is true. At least in 2003, the two highest recorded wildlife densities were for transects inside the ranch in *Orkisirata* [the early rains, December]. These numbers are truly striking, with over 370 animals per km^2 for one of the transects and 350 for the other. This represents over 1,106 zebra and wildebeest as well as other animals. And on these transects, the cattle densities were zero inside the ranch.

[*MG provides further elaboration of data. She looks back at M and K for their reaction. They seem happy, but M is anxious to say something. He is given a chance to speak directly to the transect data and he promises to be brief.*]

LANDIS M: I just wanted to speak briefly on the high densities of wildlife in the ranch that MG referred to. They were high, probably more than reported, as we had trouble counting them all! And it is true that on those same transects, there were almost no cattle present. But we need to be careful in drawing conclusions from this. First of all, wildlife [particularly zebra and wildebeest] follow rainfall, and at that time [December 2003] the ranch had received the early rains, when other areas [in the villages] did not. This is common; it is one of the reasons the ranch has always been an important grazing area for us during this time of year [*Orkisirata*]. However, now we are not allowed to take cattle to the ranch without permission and then only during the dry season, *not* during *Orkisirata*. And it is normal that during this time [October through December] there are many wildebeest and zebra throughout the area, not just in the ranch but in the village, where they graze with cattle, but in places where we did not have transects. [*He decides to end here.*]

[*An outsider speaks, an educated Maasai man working for a larger research and advocacy project, explains his excitement about the meeting—the dialogue built—and in many ways sums things up. But because he is an outsider, this does not end the meeting. The chairman speaks.*]

CHAIRMAN O: [*He stands proudly, grinning, taking responsibility in a way for the success of the meeting. He speaks confidently.*] So can we all agree that the dam should be built together with the campsite? We will need to monitor change of course with wildlife numbers and continue our requests about grazing inside the ranch if wildlife numbers increase here in the village. The transect work will help us to detect changes in wildlife numbers after the dam is constructed. We have our knowledge of where the wildlife are at different times and changes in their numbers. But sometimes our words are not taken as seriously as those of the experts [*wataalamu*]. These transect data will help us to speak together and to see how we are doing as a village supporting wildlife. At the same time, we need to make sure that wildlife provide benefits for us. "We can now start to see wildlife like our cousins . . . we can start to drink their milk."[48] Are we all in agreement?

[*There are murmurs of agreement from the crowd. No one requests to speak, a sign that they all agree.*]

Conclusion

The *enkiguena* reflects neither the modernist trope of a singular, universal reality (story) nor a postmodern celebration of plurality. The different knowledge contributions of multiple and differently situated individuals interact, overlap, and interfere with each other as they weave together (through dialogue) a consensus. As a narrative format, the *enkiguena* maintains a sense of fluidity, of motion, of knowledge as active. It is through this motion, through and around different views and knowledge, that multiplicity is more easily accommodated, woven together in a moving narrative. If differences in data formats or knowledge expressions seem incompatible (for instance, the methods or the scale of analysis do not line up) or incoherent (e.g., knowledge based on different understandings of the seasons), this does not preclude dialogue. Through dialogue, some views can become coherent to others even if they remain incompatible or disagreeable. Knowledge contributions can also remain incoherent, but at least they are now visible. By presenting various forms of knowledge in the context of an *enkiguena*, knowledge, views, and ideas that would normally be invisible become visible. Otherwise anecdotal information speaks out next to scientific reports and can even add coherence to them. Maasai knowledge and scientific knowledge are not separated out into different analyses but instead are viewed, expressed, and delivered side by side in all their multiplicity. However, it would be rare for a meeting to occur as I presented it here in that I put scientific findings, the words of powerful nongovernmental organizations, and Maasai of different social standings on the same standing, albeit with explanations of their respective power.

Yet the *enkiguena* I enacted above reflects much of the ebb and flow as practiced at a Maasai *enkiguena*. My presentation in this sense is allegorical[49] and as such is neither fact nor fiction. Or according to Haraway's (*Simians, Cyborgs*) account of the "kinship" of fact and fiction—both rooted in the experience of human action and both apparent in scientific narratives—what I present is *both* fact and fiction. Haraway (4) explains the difference as follows: "Fact seems done, unchangeable, fit only to be recorded; fiction seems always inventive, open to other possibilities, other fashionings of life." But, she also warns, "in this opening lies the threat of merely feigning, of not telling the true form of things." I avoid this threat by citing my sources and explaining the experiences behind my narrative. My story is fiction in that the meeting enacted above never actually occurred as presented. Yet my story is also fact, (re)presenting statements, knowledge claims, and numbers that *were* made, stated, and calculated through the experiences of different actors in particular times and places.[50] But again, it is fiction in forcing these facts to engage in dialogue and as such opening up other possibilities of fact. Haraway (114) suggests that creating webs of situated knowledge is dependent on dialogue, on active contestation, including the "written word." Readings, she argues, "must be engaged and produced; they do not flow naturally from the text. The most straightforward readings of any text are also situated arguments about fields of meanings and fields of power." Writing the text as a dialogue encourages just this sort of reading.

Notes

1. E. Garland, "State of Nature: Colonial Power, Neoliberal Capital, and Wildlife Management in Tanzania" (unpublished PhD diss., University of Chicago, 2006), and Roderick Neumann, "Ways of Seeing Africa: Colonial Recasting of African Society and Landscape in Serengeti National Park," *Ecumene* 2, no. 2 (1995): 149–69.

2. L. White, S. Miescher, and D. Cohen, *African Words, African Voices: Critical Practices in Oral History* (Bloomington: Indiana University Press, 2001), 4–5.

3. G. Furniss and L. Gunner, eds., *Power, Marginality and African Oral Literature* (Cambridge: Cambridge University Press, 1995).

4. Dan Brockington, *Fortress Conservation: The Preservation of the Mkomazi Game Reserve, Tanzania* (Bloomington: Indiana University Press, 2002).

5. Language translations are in Maa (the language spoken by Maasai) unless otherwise noted.

6. Julie Cruikshank, "Getting the Words Right: Perspectives on Naming and Places in Athapaskan Oral History," *Artic Anthropology* 27, no. 1 (1990): 52–65, and Julie Cruikshank, *Do Glaciers Listen? Local Knowledge, Colonial Encounters and Social Imagination* (Seattle: University of Washington Press, 2005).

7. James Clifford and George Marcus, eds. *Writing Culture: The Poetics and Politics of Ethnography* (Berkeley: University of California Press, 1986); Michel de Certeau, *The Practice of Everyday Life,* translated by S. F. Rendall (Berkeley: University of California Press, 1984); and Clifford Geertz, *Local Knowledge: Further Essays in Interpretive Anthropology* (New York: Basic Books, 1983).

8. Marlene Castellano, "Updating Aboriginal Traditions of Knowledge," in *Indigenous Knowledges in Global Contexts: Mulitple Readings of Our World,* edited by G. J. S. Dei, B. L. Hall, and D. G. Rosenberg (Toronto: University of Toronto Press, 2000).

9. See also the chapter by Hammer in this volume, which addresses (and challenges) the boundary between fictional and scientific writing styles in representing, knowing, feeling, and "making visible" experience about the world.

10. See for instance Bruno Latour, *Aramis or the Love of Technology* (Cambridge: Harvard University Press, 1996), and Sharon Traweek, "Border Crossings: Narrative Strategies in Science Studies and among Physicists in Tsukuba Science City, Japan," in *Science as Practice and Culture,* edited by A. Pickering (Chicago: University of Chicago Press, 1992).

11. A missionary priest living and working with Maasai in Kenya shared with me the following story. He was quite ill, and a group of elders paid him a visit. They asked him how he was. He responded that he was not well; in fact he was quite sick. The elders were not happy with the priest's response and told him so. They asked if he had not come there to learn from them (as he had explained he did). If that was the case, they explained, he needed to know how to respond to this question. If he wakes up in the morning and is still alive then he is well because he is alive, and so he must always respond that he is well when asked. The priest was ashamed of himself, thinking, "And I came here to teach them about God?"

12. For instance, when translating into Swahili, the word *kiongozi* ("leader") is used; when translating into English, the word *chief* is often used by outsiders as well as by Maasai themselves. Here I use the spellings used by Frans Mol, *Maa: A Dictionary of the Maasai Language and Folklore; English-Maasai* (Nairobi: Marketing and Publishing Ltd., 1972), which more clearly show the roots of the words.

13. Maasai men are socially organized into age sets that progress through a series of age grades together. When a Maasai man is circumcised (in his early teens), he joins an age set and becomes an *olmurrani.* The term *olmurrani* is often translated into English as "warrior," as *ilmurran* are recognized as the "soldiers" of Maasai society. A particular age set remains as *ilmurran* for a period of ten to fifteen years, becoming senior *ilmurran* when a new group of *ilmurran* have been circumcised and trained. They then graduate as a group to junior elders and then to senior elders.

14. Even when speaking in Maa, literate Maasai often use the Swahili word *kuhusu* when presenting an agenda item. Kuhusu means "concerning" or "in regard to" and is thus a completely neutral way of addressing a topic.

15. For instance, if the discussion is about plans for the school and an elder stands and begins talking about his cattle, there is no reason to stop him or to think that his story will not in some way, at some point, reflect on the topic at hand.

16. Donna Haraway, *Simians, Cyborgs, and Women: The Reinvention of Nature* (New York: Routledge, 1991), and Marilyn Strathern, *Partial Connections* (Savage, MD: Rowman and Littlefield, 1991), talk of "partial connections" as both the connections between different people (or groups of people) and within the same person. Both authors argue for the coexistence of multiple identities that cannot be reduced to one another but are nonetheless related (i.e., Strathern the feminist and Strathern the anthropologist). Haraway uses the trope of a cyborg to discuss partial connections that can be political (e.g., different social

identities or political commitments), material (e.g., animal-human-machine), or representational (e.g., between fact and fiction). She suggests that "a cyborg world might be about lived social and bodily realities in which people are not afraid of their joint kinship with animals and machines, not afraid of permanently partial identities and contradictory statndpoints" (154).

17. Bruno Latour, *Science in Action: How to Follow Scientists and Engineers through Society* (Cambridge: Harvard University Press, 1987), and Bruno Latour and Steve Woolgar, *Laboratory Life: The Construction of Scientific Facts,* 2nd ed. (Princeton, NJ: Princeton University Press, 1986).

18. Latour, *Science in Action;* Latour and Woolgar, *Laboratory Life;* Donna Haraway, *Modest_Witness@Second_Millennium* (New York: Routledge, 1997), 33; Sandra Harding, *Is Science Multicultural? Postcolonialisms, Feminisms, and Epistemologies* (Bloomington: Indiana University Press, 1988); David Turnbull, *Masons, Tricksters and Cartographers: Comparative Studies in the Sociology of Scientific and Indigenous Knowledge* (Amsterdam: Harwood Academic, 2000); and Andrew Pickering, ed., *Science as Practice and Culture* (Chicago: University of Chicago Press, 1992).

19. Turnbull, *Masons, Tricksters,* 210.

20. See Bruno Latour, *We Have Never Been Modern,* translated by C. Porter (Cambridge: Harvard University Press, 1993).

21. While the meetings I enact never actually occurred as I present them, the structure allows me to present different knowledge contributions in one space, in dialogue, as an expanded "method assemblage," drawing from the various methods I used during research and the various methods used by the different participants to formulate and present knowledge.

22. Landis is the name of the age set for men that were senior warriors at the start of my research and then became junior elders. Makaa were junior elders and became senior elders. Seuri and Meshuki are senior (retired) elders. *Yeiyo* is the word for "mother," and *koko* is the word for "grandmother."

23. Quotations marks are used when the exact words spoken in an interview, meeting, or text are repeated. In all other instances, such as when paraphrasing or conveying a general point spoken by many, quotation marks are not used.

24. The Manyara Ranch (MR) is a relatively new conservation area that was established by the African Wildlife Foundation (through the creation of the Tanzanian Land Conservation Trust) out of land that originally belonged to both of the concerned villages here. The MR was designed to be run as a multiple-use community-based conservation area, incorporating wildlife conservation, modern cattle ranching, and local pastoralism through dry-season grazing. See M. J. Goldman, "Strangers in Their Own Land: Maasai and

Wildlife Conservation in Northern Tanzania," *Conservation and Society* (2011), for more information.

25. Statement made at the meeting, with final sentences reflecting comments made during later conversations.

26. Conversation with a man of the Landis age set on September 30, 2003, about the construction of the dam who is echoing the comments of many others, listed here and not.

27. *Boma* is the Swahili word commonly used (even by Maasai) for the Maasai homestead, which consists of multiple houses (wives of one or multiple men who are brothers, family, or friends) and a joint kraal for cattle and others kraals for sheep and goats.

28. Statement made at the meeting.

29. Statement made at the meeting. This is the end of transcriptions from that meeting, with the remainder of the chapter drawing from a variety of sources.

30. Magadi refers to high levels of Trona, a hydrated sodium bicarbonate carbonate ($Na_3HCO_3CO_3 \cdot 2H_2O$).

31. The first part of this statement (prior to her going quiet) was from a discussion in September 2003 with a woman about the dam construction but reflects similar advice provided by other women on dam placement. The second half is constructed, reflecting the views of many different women regarding the hardships of finding water in village lands.

32. From a group interview on October 7, 2003, with several women from both villages of various age groups, although this statement reflects mostly the words of one of the eldest women from both villages. She is referring here to the fact that many villagers had migrated that year in search of water during a bad drought.

33. He was working for the AWF at the time and was in charge of the microenterprise development projects but was also head of the Kwa Kuchinja Environmental Easement Project (KEEP) project in the early stages. He has since left the AWF. This statement comes from two discussions with the author in October and November 2003. I have only changed the pronouns to reflect a conversation with Maasai rather than with me. On KEEP, see M. J. Goldman, "Constructing Connectivity? Conservation Corridors and Conservation Politics in East African Rangelands," *Annals of the Association of American Geographers* 99 (2): 335–59.

34. Many of the elders who were born here spent considerable time living and traveling in other locations, so it is possible that he speaks from an experience elsewhere. This comes from a group interview in December 2003. Most of

the words here are from the elder I describe, but others, including some junior elders, contributed as well.

35. He is referring to the government training program for game scouts in Mwanza, Tanzania.

36. Interview with game scout, October 3, 2003.

37. Interview, April, 4, 2002. The early words about the youth reflect conversations with elders throughout the study.

38. Interview, November 20, 2003.

39. See Katherine Homewood and Alan Rodgers, *Maasailand Ecology: Pastoralist Development and Wildlife Conservation in Ngorongoro, Tanzania* (New York: Cambridge University Press, 1991).

40. See R. B. B. Mwalyosi, "Influence of Livestock Grazing on Range Condition in South-West Masailand, Northern Tanzania," *Journal of Applied Ecology* 29, no. 3 (1992): 581–88.

41. Interview with game scout, September 15, 2003.

42. Discussion during a visit to the wells on August 6, 2003.

43. This last sentence is from a conversation with a different warrior from the same *boma* on August 19, 2003.

44. From the combined views of several different elders—both junior and senior—and words directly spoken at a group interview in Esilalei, December 10, 2003.

45. This last statement comes from conversations with many people in both villages on this issue. Herd boys can also control their dogs. Once out herding we saw a giraffe grazing peacefully nearby. The dog was getting ready to chase the giraffe, but I requested that the dog be held back (so I could take a picture!), and the giraffe remained unbothered.

46. MG stands for the author (Mara Goldman).

47. These words reflect observations made on the transects and comments made during the course of the study by Landis K and others. There were often wildebeest around his *boma*. At the end of 2003 when I was in Arusha, I received phone calls from people in the village telling me that large numbers of wildebeest were around all the *bomas* in Engusero. People were instructed to not go out at night for fear of lion attacks.

48. This final quote is from an earlier meeting about the campsite.

49. Allegory usually connotes a form of narrative or rhetorical strategy where the meaning is other than (more than) what is spoken. According to John Law, *After Method: Mess in Social Science Research* (New York: Routledge, 2004), allegory is also "the craft of making several things at once, what is described and what can also be read into that description" (157). Allegory in ethnography is nearly inescapable: allegory "complicates the writing and reading

of ethnographies in potentially fruitful ways" (Clifford and Marcus, *Writing Culture*, 120).

50. Again to quote Haraway (*Simians, Cyborgs,* 113), "experience, may also be re-constructed, re-membered, re-articulated."

Bibliography

Brockington, Dan. *Fortress Conservation: The Preservation of the Mkomazi Game Reserve, Tanzania.* Bloomington: Indiana University Press, 2002.

Castellano, Marlene B. "Updating Aboriginal Traditions of Knowledge." In *Indigenous Knowledges in Global Contexts: Mulitple Readings of Our World,* edited by G. J. S. Dei, B. L. Hall, and D. G. Rosenberg. Toronto: University of Toronto Press, 2000.

Clifford, James, and George Marcus, eds. *Writing Culture: The Poetics and Politics of Ethnography.* Berkeley: University of California Press, 1986.

Cruikshank, Julie. *Do Glaciers Listen? Local Knowledge, Colonial Encounters and Social Imagination.* Seattle: University of Washington Press, 2005.

———. "Getting the Words Right: Perspectives on Naming and Places in Athapaskan Oral History." *Artic Anthropology* 27, no. 1 (1990): 52–65.

de Certeau, Michel. *The Practice of Everyday Life.* Translated by S. F. Rendall. Berkeley: University of California Press, 1984.

Furniss, G., and L. Gunner, eds. *Power, Marginality and African Oral Literature.* Cambridge: Cambridge University Press, 1995.

Garland, E. "State of Nature: Colonial Power, Neoliberal Capital, and Wildlife Management in Tanzania." Unpublished PhD dissertation, University of Chicago, 2006.

Geertz, Clifford. *Local Knowledge: Further Essays in Interpretive Anthropology.* New York: Basic Books, 1983.

Goldman, M. J. "Strangers in Their Own Land: Maasai and Wildlife Conservation in Northern Tanzania." *Conservation and Society* (2011).

Haraway, Donna. *Modest_Witness@Second_Millennium.* Edited by D. Haraway. New York: Routledge, 1997.

———. *Simians, Cyborgs and Women: The Reinvention of Nature.* New York: Routledge, 1991.

Harding, Sandra. *Is Science Multicultural? Postcolonialisms, Feminisms, and Epistemologies.* Bloomington: Indiana University Press, 1988.

Homewood, K., and Alan Rodgers. *Maasailand Ecology: Pastoralist Development and Wildlife Conservation in Ngorongoro, Tanzania.* New York: Cambridge University Press, 1991.

Latour, Bruno. *Aramis or the Love of Technology.* Cambridge: Harvard University Press, 1996.

———. *Science in Action: How to Follow Scientists and Engineers through Society.* Cambridge: Harvard University Press, 1987.

———. *We Have Never Been Modern.* Translated by C. Porter. Cambridge: Harvard University Press, 1993.

Latour, Bruno, and Steve Woolgar. *Laboratory Life: The Construction of Scientific Facts.* 2nd ed. Princeton, NJ: Princeton University Press, 1986.

Law, John. *After Method: Mess in Social Science Research.* New York: Routledge, 2004.

Law, John, and Annamarie Mol. *Complexities: Social Studies of Knowledge Practices.* Durham, NC: Duke University Press, 2002.

Mol, Frans. *Maa: A Dictionary of the Maasai Language and Folklore; English-Maasai.* Nairobi: Marketing and Publishing Ltd., 1972.

Mwalyosi, R. B. B. "Influence of Livestock Grazing on Range Condition in South-West Masailand, Northern Tanzania." *Journal of Applied Ecology* 29, no. 3 (1992): 581–88.

Neumann, Roderick. "Ways of Seeing Africa: Colonial Recasting of African Society and Landscape in Serengeti National Park." *Ecumene* 2, no. 2 (1995): 149–69.

Pickering, Andrew, ed. *Science as Practice and Culture.* Chicago: University of Chicago Press, 1992.

Strathern, Marilyn. *Partial Connections.* Savage, MD: Rowman and Littlefield, 1991.

Thiong'o, Ngugi wa. *Decolonising the Mind: The Politics of Language in African Literature.* London: Heinemann, 1986.

Traweek, Sharon. "Border Crossings: Narrative Strategies in Science Studies and among Physicists in Tsukuba Science City, Japan." In *Science as Practice and Culture,* edited by A. Pickering, 429–65. Chicago: University of Chicago Press, 1992.

Turnbull, David. *Masons, Tricksters and Cartographers: Comparative Studies in the Sociology of Scientific and Indigenous Knowledge.* Amsterdam: Harwood Academic, 2000.

White, L., S. F. Miescher, and D. W. Cohen. *African Words, African Voices: Critical Practices in Oral History.* Bloomington: Indiana University Press, 2001.

Chapter 5

Sleepwalking Lands

Literature and Landscapes of Transformation in Encounters with Mia Couto

Amanda Hammar

Writing Words and Worlds

At the heart of this chapter is an exploration of the ways in which words—mostly written words but also language more broadly and fiction and poetry more specifically—engage with the world and how the world (human and nonhuman) in turn "speaks" and thus asserts and (re)creates itself. Related to this, the chapter reflects on the kinds of naturalized boundaries inherent in traditional scientific modes of knowing and representation that more intimate subjective modes can productively challenge. Such naturalized boundaries have produced *un*natural dichotomies between, for example, object and subject, nature and culture, science and art, outer and inner worlds, and the living and the dead. Confronting these dichotomies is not aimed at merging everything into indistinctness. Distinctions can and do have analytic and heuristic value. Yet the persistence of dichotomies and of the Enlightenment-based domination—perhaps even the masculinization—of science vis-à-vis humanities needs to remain constantly open to reflexive critique. This is in order to confront what Will Wright (1992) calls the "conceptual incoherence" of scientific knowledge and its various socially, politically, and ecologically legitimating and distorting projects.

My own interest here is especially in the ways in which the different worlds implied by the above-mentioned distinctions—physical worlds, worlds of imagining, spirit worlds, and worlds of practicality, production, and being—*articulate* with each other. And in this process I am curious about how the boundaries between them are altered. In examining such boundaries and the worlds they seemingly divide, my intention is to consider ways to reframe how we might read, relate to, and represent such worlds and the relationships between them.

The inspiration to consider these questions more closely came from an engagement with Mozambican writer Mia Couto's novel *Sleepwalking Land* both as a reader and in direct conversation with Couto himself.[1] These interweaving encounters with Couto and the reflections they generated on the relationships among war, violence, displacement, and environment and between language and forms of knowing and showing constitute the bulk of this chapter. But before turning directly to the substantive discussion of Couto's work and his framework for reading the world, I briefly trace my own intellectual uneasiness with the classic distinction between science and literature and then discuss how encountering Couto has added new dimensions to addressing this unease. This provokes reflections on the notion of interconnected domains of knowledge and engagement that provides a potential counterpoint to the limitations and dangers of such dichotomies.

Encountering Couto

In the academy where I was located for the period of my doctoral research in the late 1990s, particularly in the social sciences and in development studies, I was warned several times to "guard against too much language." It was a warning not to get "too carried away" with the writing itself, as though language was some kind of dangerous drug that might distort objective reality and undermine the integrity of one's analysis and the validity of one's claims. A concern with language, it was implied, was the domain of literature, of "writers" (that is, literary writers) or linguists. Although "scientific objectivity" has been an issue debated from different standpoints for many decades,[2] the narrow discourse of objective distance was nonetheless not unusual in this context, namely that social scientists should instead concern themselves with facts and evidence-based analysis, not the words (or passions) through which a social scientist translated these onto the page. The broader message was that social science and literature were somehow irreconcilably separate domains of knowledge and should remain distinct in terms of access to knowledge and claims to "truth." At the time I reacted instinctively against this seemingly unnatural separation between what we witness of the world and attempt to understand and interpret on the one hand and how we write or speak about this on the other hand. I was aware then—and have become increasingly so—that writing, be it in literary or so-called scientific forms, is never an isolated act. Language and our relationship to it are powerfully implicated in our ways of seeing, ways of knowing, ways of feeling about, and ways of representing—hence also creating—the worlds that we inhabit. In this sense, words matter deeply.

One of the spheres in which disciplinary, theoretical, and to some extent ideological divergences between science (in its environmental/conservationist form) and literature (in its postcolonial/critical theory form) have manifested is in the tension between ecocriticism and postcolonial studies. Rob Nixon has usefully summarized the specificities and "mutually constitutive silences" that have marked the divide between these two perspectives.[3] He notes that ecocritics focus more on the purity and preservation of nature, while postcolonialists are more concerned with hybridity and points of cultural intersection. The former prioritize the literature of place, and the latter prioritize displacement. Ecocriticism emerges from and reinforces a national (and primarily American) positionality, whereas postcolonialism both emanates from and is oriented toward the cosmopolitan and transnational. Finally, ecocriticism often erases histories of colonized and dispossessed peoples, while the postcolonial project is explicitly directed at "excavating or reimagining the marginalized past" (Nixon, 717).

Part of the intellectual project that underpins this volume as I understand it, and that I strongly support, is to facilitate a productive engagement between these two sets of concerns: to explore what each can offer the other as a way into a more inclusive and creative engagement with pressing questions of local and global environmental and social justice. Even if not explicitly adopting the analytical tools of either perspective here, there are certainly elements of both that inform this chapter and that are highly relevant to a reading of Couto's novel.

The disturbing disjuncture between scientific and literary sensibilities and their respective modes of engagement with the world/s we live in and envisage was the focus of a conversation with Mia Couto in Maputo in February 2008. The initial purpose of our meeting as I had envisaged it and pitched it to him was to discuss the relationship between literature and displacement and, in particular, how we could think about this in relation to

Couto's novel, *Sleepwalking Land*. I had been working for some time with questions of displacement in southern Africa (especially in post-2000 Zimbabwe and to a lesser extent from the mid-2000s also in Mozambique).[4] As such, I was constantly thinking about different forms of investigation and modes of knowing and representation around this complex, paradoxical phenomenon. Some of the fiction and poetry emerging from Zimbabwe in particular during the 2000s spoke powerfully and insightfully to the realities of state-sponsored violence and displacement in ways that social science research (politically and physically constrained as it was) was not always or entirely able to do.[5]

While Couto and I did indeed discuss the intended topic, our exchange broadened into reflections on how literature, nature, and the social world are implicated in the production of one another and with what effects. This was consistent with the themes of the novel itself, where war, violence, and displacement are inseparable from active nature and environmental change. Yet it also reflected Couto's combined orientation and practice as both a literary figure (as fiction writer and poet) and—as I was to discover during our conversation—an environmental biologist.

The novel was published in English translation as *Sleepwalking Land* in 2006 but was first published in Portuguese in 1992 with the title *Terra Sonambula*.[6] The timing of the novel's original publication is significant in that it was the same year that the conflicting parties in Mozambique's sixteen-year-long brutal civil war signed a peace agreement. However, the novel itself was written *during* the war. Among other things, as already noted, the novel engages with and is embedded within the extremes of violence, loss, and displacement—but also the paradox of creativity— that such a war entailed. As part of this, the story engages profoundly and highly imaginatively with how war and violence act upon both nature and bodies and how in turn bodies and nature (the human and nonhuman) speak their refusal of death in various tongues. In addition, Couto offers a way of thinking about the relationship among the natural, social, and spiritual worlds, between visible and invisible worlds, that engages productively with the question of boundaries.

The novel is written in a way that counters what Couto calls, somewhat disdainfully, the "functional language of science." Couto's own relationship (and challenge) to science is grounded mostly in his proximity to the natural/physical sciences as a biologist. His self-distancing from the *language* of this particular mode of knowing and representation (while not necessarily from its analytical tools, which he uses in other aspects of his work) helps sharpen the critique of its limitations. However, there are various kinds of science. My own concern is primarily with the interface between literature and *social* science (given my own training and positionality as a social scientist). Perhaps the social sciences—and especially the "softer" sciences under this umbrella, not the least of which is anthropology—allow more freedom in forms of expressing ideas and findings.[7] Nonetheless, the dry and uninspired nature of much social science writing undoubtedly contributes to its (partial) failure to connect with and convince its varied audiences. Ironically, this counters the very desire of many such scholars to communicate issues of great social (or personal) importance.[8] However, this is not an argument in favor of replacing sound evidence-based scholarship with fictional writing. Rather, it is an assertion that there is a range of creative ways in which social science can (and occasionally does) engage with and represent the real world that does not reproduce unreal dichotomies or alienate readers through abstract, disconnected language. In addition, as Lewis et al. (*Fiction of Development*) argue, fictional writing often portrays "reality" more effectively than scientific work and/or makes it more accessible to the general public and/or to decision makers.[9]

Interconnected Domains of Engagement

Central to the present discussion is the question of whether literature and science necessarily represent different *domains of engagement* (and hence knowing), be this an engagement with Couto's visible and invisible worlds or with what John Berger calls "the existent."[10] Do they necessarily do different work in terms of how they respectively approach, make sense of, and narrate their subject matter? Do they necessarily draw on different kinds of sensibilities or skills that then in turn open or close certain windows of understanding and analysis? Can we look meaningfully, productively, through several windows simultaneously, and if so with what effects?[11]

In my initial thinking about these questions for the purpose of this chapter, I became conscious of an implicit divide that I had internalized whereby I thought of fiction or poetry as being the *realm of emotion and of intuition* and thought of science or social science as *the realm of precision and explanation*. Both necessarily had some kind of relationship to "the empirical" in that I saw them as both (potentially at least) drawing from lived realities and experience, or engagement with others' lived worlds, and I saw each as contributing to expanded knowledge or insight. Yet there was clearly a sense that each domain was defined by different and distinct "rules" of access to experience/knowledge and its wider, more public, translation. As already discussed, this dichotomy is powerfully embedded in both intellectual and institutional practices within the academy as well as the policy world. As such, one of the points of this discussion was to counter the analytic limitations and even the potentially dangerous distortions of this divide.

Once, Couto muses, science and literature were equally realms of wonder, curiosity, and passion. The divide between them, however, that now seems so automatic wasn't always given. According to historian Hayden White (1973), there was not much difference in the structuring of nineteenth-century European historical accounts and realist novels of the time.[12] Both, he argued, derived their "force and persuasion" primarily from the deployment of rhetorical strategies. Resorting to figurative rather than objective forms of representation was a particularly apt persuasive device when facts were not directly available.[13] But for Couto, relating to contemporary times, it is primarily poetry (in its broad sense)—"the language of the dreaming world," as he calls it—that allows us to connect to "things that we can't know using the formal, conventional knowledges and languages." He claims that it is poetry (and by extension, fiction) that gives us access to "the underground world." Couto's vision as a writer is to "give back to the word its divine power," its power to create rather than merely nominate things, the power to enchant things, be these trees, birds, or landscapes. *Sleepwalking Land* is written in just such a way. In this sense it asserts the power of writing to transform the world.

Science, in Couto's experience, "has become a very accommodative thing; more a function." As a scientist, he suggests, "you are serving certain powers and you become a kind of functionary of these interests. So science has become something that is not really scientific, [through losing] this desire, this passion of knowing something that is not certain." This returns us to the question of whether there are differences between the natural or physical sciences and the social sciences. Are these claims of Couto's as valid for the latter as the former? Are social scientists, similarly, mere functionaries of more crudely narrow interests and just as creatively or analytically constrained by scientific rules concerning method, form, and language? Again, I would argue that the distance from literature and its possibilities is less sharp among at least certain social sciences or certain social science practices. Therefore, while there is a fair degree of empirical truth to Couto's general suggestion that science often undermines or delegitimizes passion, I would nonetheless counter any simplistic generalization about science or scientists per se being unimaginative or unable to cross from the "realm of precision and explanation" into "the realm of emotion and intuition." As already noted, Couto's version of science draws most directly from his combining of literary writing with work as an environmental biologist, and perhaps herein lies one source of his own creative tension. Yet *his* science, I would suggest, is *that* science.[14] And there are other kinds of sciences, so that even while I see his overall critique as valid and useful, one of the strongest *counterarguments* to a too-simplistic generalization about "science" (or bad science) can be drawn from good (and well-written) ethnography.

Harry West's beautifully crafted ethnographic study of Mozambique's postwar dynamics of "governance and the invisible realm," *Kupilikula* (2005), is a case in point.[15] Much as Couto in *Sleepwalking Land* approaches the unknown fearlessly (yet not without very controlled literary craft) and challenges the false divides between "nature" and "culture," visible and invisible worlds, "the living" and "the dead," West attends intimately to the ambivalent boundaries and relationships between the living and the dead, in this case in the aftermath of Mozambique's civil war. There is a sense of West's ability and willingness to enter the worlds he is investigating with a deep passion for, as Couto puts it, "knowing something that is not certain." West's work on sorcery practiced by Muedan healers in northern Mozambique addresses among other things the "invisible realms" produced and inhabited by sorcerers that allow them to gain powerful perspective on the visible. West too speaks of his own work of "making the invisible visible." As ethnographer, through "ethnographic sorcery," he makes "of Mueda and Muedans something that they themselves have not; I have made them in accordance with my own vision" (83). West's close ethnographic encounters were in part a journey "into the realm of hidden, but decisive, forces" (79). He has dared to "go in close" (Berger, *Shape of a Pocket*, 16). It is therefore not only poetry and fiction, as Couto proposes, that gives access to "the underground world." But it might be that it is the poetic wonder, the suspension of fixed and restrictive analytic boundaries, that allows West to cross between these seemingly distinct but clearly so connected worlds.

Following from the discussion above, I would argue further here that *intimacy* and *visibility* are not only two critical, complementary dimensions of "the desire to know something that isn't certain," that isn't easy to grasp, but also important tools in analyzing the differences as well as the links between these apparently dichotomous forms of investigation, recording, and representation. By *intimacy* here I mean mostly connection or connectedness and being moved; by *visibility* I mean both seeing or witnessing and making visible.

A concern with these interrelated sensibilities/practices poses an ongoing challenge to my own double-sided engagement with questions of displacement and dispossession. On the one hand, as indicated earlier, I have worked with such issues as a researcher and an academic since the late 1990s.[16] On the other hand, these are themes that have engaged me personally, politically, and creatively for many decades and are informed by my biography: as the grand-daughter of Jewish immigrants in southern Africa who fled the pogroms in eastern Europe in the 1890s, as a Zimbabwean deeply conscious of the colonial dispossessions that shaped my country, and more recently as witness to the postindependence violence and displacements that have reshaped Zimbabwe again and that have in turn shaped and sustained my own partial exile. These are themes that are part of the everyday material with which I construct my own belonging in strange places and that get expressed also through nonacademic writing: in personal journals, in commentaries on the situation in Zimbabwe, and occasionally in poetry.

On the surface the academic and the personal seem like different domains and even scales of engagement and knowledge production—"scientific" research and creative writing—in the sense of seeming like different realms of sensing, seeing, and translating the experiences and effects of displacement. And perhaps to some extent they are necessarily distinct but not entirely so. Certainly each is both limited and liberated by its own discursive parameters and boundaries of interpretation and communication, related partially to assumed or actual audiences. Yet the connective tissue between them is precisely *connectedness.* In both cases what is critical is my proximity to the issues at stake (intellectually, politically, personally) and additionally—by consciously remaining open to such intimacy—being moved to engage, to witness, to record, and to make visible.

However, even while promoting close connection with one's subjects of research (or in the case of art, with the artist's model), John Berger (*Shape of a Pocket,* 16) warns that there are risks involved in the intimacy of connection: "To go in close means forgetting convention, reputation, reasoning, hierarchies and self. It also means risking incoherence, even madness. For it can happen that one gets too close and then the collaboration breaks down and the painter [or writer] dissolves into the model [or subject]." For Couto, but also for social scientists committed to what I would term *intimate modes of knowing,* such risks seem worthwhile. They are the risks associated with integrity and with the necessary passion of poetry and fiction. This is not to suggest that such literary forms are without convention. But when compared with the strict conventions of "scientific" writing, even if imaginatively crafted, novels or poetry can more easily cross between worlds. And importantly, in doing so they can more easily represent *the simultaneity of worlds.*[17] They can legitimately interrupt commonsense, linear notions of space and time through what Ranka Primorac calls multiple "novelistic possible worlds."[18] Couto does this particularly well in *Sleepwalking Land.*

Dissolving Boundaries

I turn now to the novel. Through stories within a story—in which writing itself plays an important part—*Sleepwalking Land* traces several kinds of interweaving journeys through the war-ravaged landscapes of Mozambique. These are landscapes in which the boundaries between animals and humans, between land and sea, between the living and the dead, frequently dissolve.

The novel begins with two refugees—an old man, Tuahir, and a young boy, Muidinga—walking along a road in a landscape that "had blended sadnesses the likes of which had never been seen before, in colours that clung to the inside of the mouth. They were dirty colours, so dirty that they lost all their freshness, no longer daring to rise into the blue on the wing. Here the sky had become unimagineable. And creatures had got used to the ground in resigned apprenticeship of death" (Couto, *Sleepwalking*, 1). They have left a refugee camp. The two, writes Couto, "matched the road, withered and devoid of hope" (1). They have no clear destination: "Their destination is the other side of nowhere, their arrival a non-departure, awaiting what lies ahead" (1). But even if they seem to be in search of another place in Mozambique, as Couto noted in our interview, this was not a geographic place. It was something else, "a hidden place," he said. At the same time, the notebooks of a young man called Kindzu, which they find in a suitcase beside an unidentified body close to a burned-out bus on the side of the road and which Muidinga reads to Tuahir during the course of the novel, speak of a journey to other places, both interior and exterior.

Already in the first paragraphs we get the sense of nature being both observer and active creator of the world: beyond the dead road, beyond the burned-out bus, "only the baobabs contemplate the world shedding its flowers" (1). Throughout the book there are trees or birds or other animals speaking, both menacingly and gently, as ghouls or guides on Kindzu's uneven journey in search of the *naparama* ("warriors of justice") in the north, whom he hopes to join. There are spirits that appear and disappear, dwarfs that drop from the sky, and hands that spring from the soil. In many senses, nature keeps hindering human endeavor. Kindzu must *become* nature in order to endure it. He must heed the words of the *n'anga* (shaman or traditional healer) whose advice he sought before beginning his journey: "in the sea, you will become the sea," Kindzu had been told, and when at one point he has to row with his bare hands after losing his oars, his fingers grow webs. "In the water," he writes in his notebook, "I felt I had scales instead of skin" (36).

Much of Couto's imagery of Mozambique in a time of war is of people no longer themselves, losing a sense of who they were. As Kindzu says early on in his tale: "Little by little we became different, unrecognizable" (10). Similarly, the landscape is no longer itself; it is turned upside down. Here it is Couto the environmental biologist commenting on how the war has altered people to the point that they dangerously lose respect for nature. He describes in one passage how "the sea dried up completely and all the water disappeared within an instant." In another passage he describes coconuts looking like "golden gourds containing a thousand riches," but then a voice is heard from one of the trees "begging the men to pause and ponder" before stripping them of their fruit, "for the destiny of our world was held together by delicate threads" that if severed would precipitate "an outbreak of disorder and a whole succession of disasters" (13). So, for example, the first fruit is cut, and a flood occurs.

The metaphor here and elsewhere seems to be a commentary on the destruction of the fruits of liberation through a senseless war but also through greed. As Kindzu observes at one point on his journey, "Now, I saw my country like one of those whales that come to breathe their last on the shore. Death hadn't even occurred and knives were already stealing chunks of it, each one trying to get a bigger piece for himself. As if it were the last animal, the last chance to gain a share. From time to time, I thought I could still hear the giant sighing, swallowing wave after wave, turning hope into an ebbing tide" (16). Occasionally there are moments of hope, but these are usually fleeting, and Couto's own environmental consciousness can be heard making a warning link between survival and respect for and understanding of nature. At one point Muidinga senses hope returning to his bleak surroundings, but this can only be sustained through a deeper connectedness to nature and its wisdoms by humans: "Muidinga notices that the features of the landscape around them are changing. The earth is still parched, but among the sparse tufts of grass hang the remains of the morning mist. For Muidinga, the moisture is a portent of green. It's as if the earth is waiting for villages, dwellings, to nurture dreams of future happiness. But the untamed bush doesn't offer sustenance to those who are ignorant of its secrets" (45).

While Kindzu's journey is one of constant movement and adventure and engagement with a combination of people, nature, and spirits,

Muidinga and Tuahir barely move from the bus that they have made their temporary home. Instead, the landscape around them shifts constantly, allowing no certainty. When they do venture out, they too encounter different kinds of strange characters. On one of their sorties for food, they are captured by the one-eyed old man, Skellington, who has remained alone in his village while everyone else has fled. He explains how he has survived all the pillaging, killing, and burning: "I'm like a tree," he says. "I just pretend to die" (63). He is old and very tired but stays alive in order to guard his village. Muidinga has felt offended by Skellington taking Tuahir and himself prisoner and reprimands the old man for his lack of hospitality, for not receiving them as visitors as custom dictates. But in these times the world is not as it was. Tuahir, usually gruff, offers the old man words of hope: "Our land will quieten down, all will become kin, all Mozambicans. And we will visit each other as in times gone by, gnawing the road without fear anymore" (64). (Interestingly, this is one of the few passages in which an explicit reference is made to the nation and a vision of a revived national belonging.)

But despite Tuahir's comforting words, Skellington keeps the two of them tied up in a net as prisoners. During the night the young boy manages to free one of his arms and, with a stick, writes something on the ground. The next morning when Skellington wakes and goes to them and sees this, he asks, "What are those drawings?" "It's your name," Tuahir replies. "Is that my name?" Skellington asks in wonder, unable to read. There is a transformation in the old man. He hums and chants and smiles. Muidinga is lulled to sleep but later awakens to find Skellington standing over him with a knife. The old man releases his two prisoners and leads them through the bush for a long time until they reach a large tree. There Skellington indicates to Muidinga to write the old man's name on the tree with the dagger.

> Muidinga carves the letters of Skellington's name in the trunk. . . . The old man passes his hand over the bark of the tree in rapture. Then he says: "You can go now. The village will survive, my name is in the blood of this tree now." Then he puts his finger in his ear, inserting it deeper and deeper until they hear a muffled sound of something bursting. The old man extracts his finger and his ear spurts a fountain of blood. Gradually, he wastes away until he is no more than the size of a seed (67).

This is one of several ways in the novel in which writing, or in this case a single word, has that power that Couto spoke of to enchant, to create, and to transform the world.[19] At the same time, Couto evokes here the sense of language *and* nature simultaneously creating the world. Elsewhere the young woman, Farida, writes letters of an imagined past for the older woman who has adopted her to help her adopted mother live in the unbearable present. Writing in the novel (and perhaps writing the novel itself, for Couto) is about making the invisible visible and through this writing one's way both physically and metaphorically to somewhere else. Writing is almost always related in the novel, in some way, to transformation. Kindzu's own writing of the notebooks, for example, keeps him alive, keeps him dreaming of another place, a future, even as what he writes, as he says toward the end, "depends on what I'm dreaming" (190).

Sleepwalking Lands

The novel is, as we know, called *Sleepwalking Land,* or *Terra Sonambula.* As previously noted, somnambulate is defined as "to walk or perform another act while asleep or in a sleeplike condition."[20] Not surprisingly, then, the story is rich in metaphors of sleeping and dreaming. *Sleepwalking Land* reveals a land that itself moves while people sleep, moving inside and through their dreaming. Again, what this points to is land or nature as *active,* itself producing the world, in a dynamic yet hidden relationship with the characters of the novel, echoing Donald Moore's assemblages or "natural-cultural hybrids," mixed entities shaping power and place.[21] Land, here, is not simply "the inescapably solid, tangible and material" form of "natural landscape scenery" traditionally associated with the notion of landscape.[22] Rather, land is simultaneously the imagined landscape of an interior world *and* terrestrial or anthropological space transformed "into a place of historical life."[23] In other words, interior and exterior landscapes are mutually constitutive.

But the notion of sleepwalking land is also about the blindness and denials, the closed-eyedness, and the constrained consciousness produced by and producing the violence, displacements, and environmental destruction of war. This notion speaks of a blind wandering in search of hidden places or invisible worlds, a search for belonging, and a search for meaning in the face of a mad logic of war that, for Couto, "had contaminated the whole country." There seems to be much attention given in postcolonial literary studies to the significance of landscape in the construction of postcolonial identities and nations.[24] But Couto's project in *Sleepwalking Land* does not seem to be about that. Instead, it is more pointedly about the relationship between the visible and invisible worlds that we inhabit and create. In a context of war and displacement, inevitably there are questions about—and even a quest for—that which remains solid in a profoundly shifting and uncertain environment. In these moments—extended moments in the case of Mozambique's long civil war—of overwhelming loss and turbulence, of literal and metaphorical displacements, not only does the issue of belonging come to the fore, but so too does even the basic question of *being* at all. In a way, the book appears to be Couto's own way of figuring out how to *be* in the face of his country's interminable, brutal war and its multilayered dislocation and destruction of both humans and nonhumans.

In our interview in February 2008, I asked Couto why he had written this particular book when he had. The first time I asked this, he answered as follows:

> That book, it's the only book I have suffered to write. Usually, I write for pleasure. But that one, it was especially [difficult]. It was during the war. The last two years of the war. And I had this feeling that it was impossible to write about the war during it. It was something so cruel, so close to us that I thought we would only be able to write about the war after the war. But it happened like this. I don't like to romanticize the act of writing. But in that case I was being visited by voices and by stories. I don't believe in God but I was almost praying to be separated from these voices, because I really needed to sleep. It was sleepless nights. And then I realized it was some feelings, some signals of something happening. The peace agreement was being prepared, but I didn't know. So maybe I was receiving [signals] not because I have special feelings but because . . . [there was a different energy in the air]. But we couldn't believe that it would be the end of the war, because for us, that war would not end. That was the feeling.

Later in the conversation I asked if writing the novel was a way of being able to speak about the unbearable. "I think there are always things that are unbearable," Couto began. "It's not just a question of speaking about a special time or special reality." Delaying an answer about his own situation, he spoke about Ho Chi Minh, who had written a book of poetry while in prison. On being asked how he could write such beauty in conditions of great difficulty, Ho had apparently answered (in Couto's translation), "I devaluated the walls," meaning that he deprived them of their value; he removed their weight. It was clear that Couto related strongly to this act of devaluating that which is too much to bear. And so I pushed a little harder. Did he write the book, I asked again, because there was no way of talking directly about the horrors of the war? "Yes, that's right," he began, then continued:

> I think this brings us to this question of displacement, in terms of war; to resist against this *total* sense of displacement that is propagated by the war—maybe not just the war, but *sixteen years* of war; a war that doesn't fit with anything logical. You know, all the analyses that you could produce don't explain anything. Something [is] always [flowing invisibly] underground. And then you have this impression that [in order] to be in touch with this untouchable thing, you should *choose* your displacement. It should be owned by you, this other [interior] displacement. You know, [you say to yourself] I'll be in this no-place. That's what the narrator of the book does. He's looking for some other place in Mozambique. In principle, at the surface, it seems that he's just looking for some place to survive, . . . [yet it's] some other place which is not a geographic place but an invented place [that] can only be born in the border between the written and the oral world.

Here we diverged again slightly and talked for a while about how poetry in particular can access and reveal the underground world, how it isn't bound by a linear narrative. But still I felt compelled to dig deeper for his story of writing himself to somewhere else. And so I asked again if writing the book helped him to find a way of being within a place of such uncertainty.

> Yes, yes. You see there were friends that I lost during the war, very close friends. And during that time, I was not able to remember. It was very strange. I, I was afraid . . . to write that book . . . but I'm not sure if it was the book, to tell you the truth. . . . Maybe it was both, [also] the [other] thing. Because as I told you, during the time that I was writing the book, I think it was two months after I finished my story there was a peace agreement, so there was something more real and more social that was happening at the same time. But then I realized during this time, during this period, that I was beginning to be able [to remember] those persons again. I had buried [things] . . . not in terms of buried to forget, but I created my own past. I did it because, you know, [those] persons, those dead persons, were

never finishing to die. They were dying everyday. So, after the book, they become quiet. And they become really dead. And I could visit them. And they could visit me.

Conclusion: Writing the Unbearable into Hope

In going to this place of intimacy within his own writing—the intimacy that liberates in making itself visible through writing—I finally understood why, for me, Couto's novel has worked so powerfully. In *Sleepwalking Land*, Couto "goes in close" fearlessly and succeeds in making visible to us invisible worlds as well as the relationship between the visible and the invisible. He does this because he is committed to "the language of dreaming," the language of intimate connection to the worlds of the interior and the hidden. Others who are not fiction writers occasionally do that too and do it well, while there are countless examples of both fiction and nonfiction writing that fail to move the reader and hence fail to do their work. With scientific writing in particular, there is the danger of being constrained by the formalism of method and/or language that reinforces a distance from the subjects and subject matter of research and retains the safety but also the emptiness of abstraction. The point here is not to promote sentimentalism in relation to that which is unbearable, nor is it to encourage idealizing the creative agency of those who survive and transform what seems impossible to bear. Rather, it is to ensure a greater coherence in our knowledge and representation of the interconnected social and natural worlds that we inhabit, study, and coproduce (with both humans and nonhumans). It is to encourage ourselves to do so as boldly and honestly as possible, to dare us to "go in close," at least close enough, to witness and recount the complexities and contradictions and the breaking of boundaries that our empirical encounters with multiple realities demand.

The argument beneath this is that forms of knowing and writing that are more aligned with the principles of poetry and fiction, rather than those constrained by the formalized rules of traditional science, can and do help us to "go in close(r)" and to achieve greater coherence in our understandings of the relationship between the social and natural worlds. However, much of the success of such an endeavor rests on the nature and quality of translation from observing and engaging to knowing and from knowing to representation. The quality of writing itself, I would argue, is related at least in part to the degree of openness and intimacy that each researcher has allowed space for during both the research process and the writing process. "Translation is a kind of transubstantiation," writes Anne Michaels, referring to poetry, but I would suggest that this is true with much wider relevance.[25] Researchers as well as poets and fiction writers are in some senses translators of life into language. "You can choose your philosophy of translation," Michaels continues, "just as you choose how to live: the free adaptation that sacrifices detail to meaning, the strict crib that sacrifices meaning to exactitude." Either way, the job demands that we "try to identify the invisible, what's between the lines, the mysterious implications" (*Fugitive Pieces*, 109). For now, I conclude with a simple supposition: that greater connectedness to the things we choose to write about—whether this is written as fiction or faction—and a commitment to making visible the uneasy paradoxes of "truth" and the awkward simultaneities of being open up a more honest, revealing, and meaningful engagement with both the unbearable and transformative dimensions of the world that we inhabit and attempt to understand. Such connectedness contributes to what Berger (*Shape of a Pocket*, 22) calls "an act of resistance instigating hope." In working with questions of violence and displacement and with threats of environmental destruction and injustice, this is something worth considering.

Notes

1. Preempting this was the invitation to participate in the colloquium "Environment and Literature in Africa" at Kansas University in March 2008, which underpins this volume. My sincere thanks to Garth Myers and Byron Caminero-Santangelo for the opportunity to engage with them and with all the participants at the colloquium on this important and evocative topic. I would also like to thank Ross Parsons and the two anonymous reviewers for insightful comments on an earlier draft of this essay.

2. For feminist-inspired perspectives on this question, see Donna Haraway, "Situated Knowledges: The Science Question in Feminism and the Privilege of Partial Perspective," *Feminist Studies* 14, no. 3 (1988): 575–99, and Sandra Harding, *The Science Question in Feminism* (Milton Keynes: Open University Press, 1986). On the challenges posed by science studies more generally, see Bruno Latour, *Pandora's Hope: Essays on the Reality of Science Studies* (Cambridge: Harvard University Press, 1999).

3. Rob Nixon, "Environmentalism and Postcolonialism," in *Postcolonial Studies and Beyond*, edited by Ania Loomba, Suvir Kaul, Matti Bunzl, Antoinette Burton, and Jed Esty, 233–51 (Durham, NC: Duke University Press, 2005).

4. At the time I was coordinating a collaborative research program at the Nordic Africa Institute (Uppsala, Sweden) titled "Political Economies of Displacement in Southern Africa."

5. For a selection of relevant novels, see, for example, Valerie Tagwira, *The Uncertainty of Hope* (Harare: Weaver, 2007); Brian Chikwava, *Harare North* (London: Jonathan Cape, 2009); and John Eppel, *Absent: The English Teacher* (Johannesburg: Jacana, 2009). For short stories, see Irene Staunton, ed., *Writing Still* (Harare: Weaver, 2003); Chris Mlalazi, *Dancing with Life: Tales from the Township* (Bulawayo: 'amabooks, 2008); and Pettina Gappah, *An Elegy for Easterly* (London: Faber and Faber, 2009). For a selection of poetry, see the Zimbabwe page "Poetry International" at http://zimbabwe.poetryinternationalweb.org/ and Tinashe Mushakavanhu and David Nettleingham, eds., *State of the Nation: Contemporary Zimbabwean Poetry* (Faversham: Conversation PaperPress, 2009).

6. Somnambulate: "To walk or perform another act while asleep or in a sleep-like condition." Source: Dictionary.com, *The American Heritage® Dictionary of the English Language, Fourth Edition* (Boston: Houghton Mifflin, 2004), http://dictionary.reference.com/browse/somnambular (accessed March 27, 2008).

7. James Clifford and George E. Marcus, eds., *Writing Culture: The Poetics and Politics of Ethnography* (Berkeley and Los Angeles: University of California Press, 1986).

8. On the value and effectiveness of fictional representations of key development issues, see David Lewis, Dennis Rodgers, and Michael Woolcock, "The Fiction of Development: Knowledge, Authority and Representation," Working Paper 05-61, London School of Economics and Political Science, 2005, available at LSE Research Online, eprints.lse.ac.uk/379/.

9. See Mara Goldman, this volume, for interesting reflections on the dynamic relationship between various knowledges and diverse forms of interpretation and expression, illustrated through a discussion and application of the Masaai *enkiguena* form of open-ended exchange.

10. John Berger, *The Shape of a Pocket* (New York: Pantheon, 2001).

11. Couto interview, February 20, 2008: "I'm not saying I discredit science. Science or biology is just a window. But I have more windows to look through." Unless otherwise referenced with respect to *Sleepwalking Land,* all quotes from Couto are drawn from the interview in Maputo in February 2008.

12. Hayden White, *Metahistory: The Historical Imagination in Nineteenth-Century Europe* (Baltimore and London: John Hopkins University Press, 1973).

13. This discussion draws directly from Lewis et al., "Fiction of Development," 3.

14. This is not to suggest by any means that there aren't countless ways in which the social sciences haven't dutifully served power, not least colonial power. However, that is not the point of the discussion here. Rather, we are talking about different "windows" into or access to "the invisible worlds" that require more open and less formalistic and functionalist paradigms and tools.

15. Harry West, *Kupilikula: Governance and the Invisible Realm in Mozambique* (Chicago and London: University of Chicago Press, 2005). For another example of creative social science that attends to "the invisible," in this case making use of photography as well as text, see Filip de Boeck and Marie-Francoise Plissart, *Kinshasa: Tales of the Invisible City* (Gent: Ludion, 2004).

16. For work in this mode in different contexts and periods, see, for example, Amanda Hammar, "'The Day of Burning': Eviction and Reinvention in the Margins of Northwest Zimbabwe," *Journal of Agrarian Change* 1, no. 4 (2001): 550–74, and Amanda Hammar, "Ambivalent Mobilities: Zimbabwean Commercial Farmers in Mozambique," *Journal of Southern African Studies* 36, no. 2 (2010): 395–416.

17. Doreen Massey, "Is the World Getting Larger or Smaller?," essay published on openDemocracy, February 15, 2007, http://www.opendemocracy.net/globalization-vision_reflections/world_small_4354.jsp (accessed March 23, 2008).

18. Ranka Primorac, *The Place of Tears: The Novel and Politics in Modern Zimbabwe* (London and New York: Tauris Academic Studies, 2006), 56.

19. There is an implicit link here to the question of literacy and its liberating potential both for individuals and the nation, revealing yet another side of Couto's critical consciousness woven into the novel.

20. See note 6.

21. Donald Moore, *Suffering for Territory: Race, Place and Power in Zimbabwe* (Durham, NC, and London: Duke University Press, 2005), 23.

22. Magali Compan and Katarzyna Pieprzak, "Introduction: Remapping Uncertain Territories: Towards a New Francographie of Landscape," in *Land and Landscape in Francographic Literature: Remapping Uncertain Territories*, edited by Magali Compan and Katarzyna Pieprzak (Newcastle, UK: Cambridge Scholars Publishing, 2007); and Terence Ranger, *Voices from the Rocks: Nature, Culture and History in the Matopos Hills of Zimbabwe* (Harare: Baobab; Bloomington and Indianapolis: Indiana University Press; Oxford, UK: James Currey, 1999).

23. Homi Bhabha, *The Location of Culture* (London: Routledge, 1994), 143.

24. Bhabha, *Location of Culture*. See also David Johnson, "Literatures of Nation and Migration: Charles Mungoshi, Nadine Gordimer, and the Postcolonial," unpublished manuscript, n.d.

25. Anne Michaels, *Fugitive Pieces* (New York: Vintage Books, 1997), 109.

Bibliography

Berger, John. *The Shape of a Pocket.* New York: Pantheon, 2001.
Bhabha, Homi. *The Location of Culture.* London: Routledge, 1994.

Clifford, James, and George E. Marcus, eds. *Writing Culture: The Poetics and Politics of Ethnography.* Berkeley and Los Angeles: University of California Press, 1986.

Compan, Magali, and Katarzyna Pieprzak. "Introduction: Remapping Uncertain Territories; Towards a New Francographie of Landscape." In *Land and Landscape in Francographic Literature Remapping Uncertain Territories,* edited by Magali Compan and Katarzyna Pieprzak, 1–9. Newcastle: Cambridge Scholars Publishing, 2007.

Couto, Mia. *Sleepwalking Land.* Translated by David Brookshaw. London: Serpent's Tail, 2006.

De Boeck, Filip, and Marie-Francoise Plissart. *Kinshasa: Tales of the Invisible City.* Gent: Ludion, 2004

Hammar, Amanda. "Ambivalent Mobilities: Zimbabwean Commercial Farmers in Mozambique." *Journal of Southern African Studies* 36, no. 2 (2010): 395–416.

———. "'The Day of Burning': Eviction and Reinvention in the Margins of Northwest Zimbabwe." *Journal of Agrarian Change* 1, no. 4 (2001): 550–74.

Haraway, Donna. "Situated Knowledges: The Science Question in Feminism and the Privilege of Partial Perspective." *Feminist Studies* 14, no. 3 (1988): 575–99.

Harding, Sandra. *The Science Question in Feminism.* Milton Keynes: Open University Press, 1986.

Johnson, David. "Literatures of Nation and Migration: Charles Mungoshi, Nadine Gordimer, and the Post-colonial." Unpublished manuscript, n.d.

Latour, Bruno. *Pandora's Hope: Essays on the Reality of Science Studies.* Cambridge: Harvard University Press, 1999.

Lewis, David, Dennis Rodgers, and Michael Woolcock. "The Fiction of Development: Knowledge, Authority and Representation." Working Paper 05-61, London School of Economics and Political Science, 2005. Available at LSE Research Online, eprints.lse.ac.uk/379/.

Massey, Doreen. "Is the World Getting Larger or Smaller?" Essay Published on openDemocracy, February 15, 2007, http://www.opendemocracy.net/globalization-vision_reflections/world_small_4354.jsp (accessed March 23, 2008).

Michaels, Anne. *Fugitive Pieces.* New York: Vintage Books, 1997.

Moore, Donald. *Suffering for Territory: Race, Place and Power in Zimbabwe.* Durham, NC: Duke University Press, 2005.

Nixon, Rob. "Environmentalism and Postcolonialism." In *Postcolonial Studies and Beyond,* edited by Ania Loomba, Suvir Kaul, Matti Bunzl,

Antoinette Burton, and Jed Esty, 233–51. Durham, NC: Duke University Press, 2005.

Primorac, Ranka. *The Place of Tears: The Novel and Politics in Modern Zimbabwe*. London: Tauris Academic Studies, 2006.

Ranger, Terence. *Voices from the Rocks: Nature, Culture and History in the Matopos Hills of Zimbabwe*. Bloomington: Indiana University Press, 1999.

West, Harry G. *Kupilikula: Governance and the Invisible Realm in Mozambique*. Chicago and London: University of Chicago Press, 2005.

White, Hayden. *Metahistory: The Historical Imagination in Nineteenth-Century Europe*. Baltimore: John Hopkins University Press, 1973.

Wright, Will. *Wild Knowledge: Science, Language, and Social Life in a Fragile Environment*. Minneapolis: University of Minnesota Press, 1992.

Chapter 6

No Longer Praying on Borrowed Wine

Agroforestry and Food Sovereignty in Ben Okri's
Famished Road Trilogy

Jonathan Highfield

IN *THE Wretched of the Earth*, Frantz Fanon writes that "The relations of man with matter, with the world outside, and with history are in the colonial period simply relations with food."[1] By this Fanon means that existence itself is so threatened that every bit of food that a person can gain access to is "a victory felt as a triumph for life" (308). He insists that "Independence is not a word which can be used as an exorcism, but an indispensable condition for the existence of men and women who are truly liberated, in other words who are truly masters of all the material means which make possible the radical transformation of society" (310). In the context of the essay in *Wretched of the Earth*, Fanon is clear that control over food and food production is an integral part of the process of decolonization. Explicitly echoing Fanon, South African minister of agriculture, conservation, and environment Khabisi Mosunkutu has called for the move away from "being Europe's small farmers who specialise in unfinished products"[2] toward agricultural policies that will "increase our capacity to deliver on the mandate to give all our people access to sufficient food."[3] Mosunkutu's speech is a clear reminder that food policies after independence often resemble the policies created during the colonial era. In order to generate income and support the European economy, colonial authorities insisted that the cultivation of cash crops take precedence over the cultivation of local food crops.[4] Even today, much agricultural production in formerly colonized regions focuses on inexpensive high-output starches to feed the region's inhabitants and high-protein crops and livestock produced for export.[5] Such bifurcation of agriculture means that the variety of crops grown has diminished. As Hope Shand of the Action Group on Erosion, Technology and Concentration has pointed out, "The loss of cultural diversity is intricately linked to the loss of agricultural biodiversity."[6] The aphorism "Tell me your stories and I will tell you who you are" and Anthelme Brillat-Savarin's response "Tell me what you eat, and I will tell you what you are" both speak to the importance of folkways and foodways for identity. As globalization changes the relationship of people across Africa with the food they produce and consume, identities also change.

Ben Okri's *Famished Road* trilogy (*The Famished Road*, 1991; *Songs of Enchantment*, 1993; and *Infinite Riches*, 1998) explores changing identities as global capitalism and imposed consumer culture impact the lives of the characters depicted in the novels. While the frequent use of Yoruba terms suggests that the novels take place in what is now southwestern Nigeria or southeastern Benin, Okri seems to be writing allegorically about a multivalent West Africa.[7] Initially in *The Famished Road* the time period is similarly indefinite, although as the trilogy progresses there are indications that the narratives are set in Nigeria in 1960.[8]

The Famished Road trilogy centers on transitions; the action dwells often in the interstitial space between the spirit world and the human world, between colonialism and freedom, between forest and city, between hunger and satiation. Azaro, the narrator, is an *abiku*, a spirit child who decides to stay with his parents instead of returning to the Land of Beginnings. Because he has broken a pact with his spirit-companions, they haunt him throughout the novel, trying to convince or trick him into returning with them to the realm of the spirits. The two main intersection points between the spirit and human worlds are the road and the forest. Azaro's spirit companions often lure him to one of these two places, and the trilogy suggests that there is an important dichotomy between them.

Forests and the loss of them are a central theme in the *Famished Road* trilogy, not in small part because their loss signals a significant shift in the economy and culture of the region. When Azaro's father introduces him to the world of the forest, it becomes clear that the forest in Okri's novels is a mixed-use space, a space of nature and culture, agroforestry existing within a larger and more lightly managed space. In *The Famished Road* the forest is used by farmers, by herbalists, and by hunters. The forest also contains spirits, most of whom only Azaro can see. Even as he introduces his son to the forest, Dad indicates that despite its fecundity, the forest will vanish: "Sooner than you think there won't be one tree standing. There will be no forest left at all. And there will be wretched houses all over the place. This is where the poor people will live" (34). Okri establishes a difference between the way the forest is managed by locals for the demands of human culture—food, medicine, shelter—and the way the foreign engineers direct its removal in order to open up the culture to neocolonial exploitation. In a telling passage from *The Famished Road* in the "future present," Azaro watches "ghostly wood-cutters axing down the titanic irokos, the giant baobabs, the rubber trees and obeches." The loss of potential in the cutting down of these trees is revealed symbolically as the nests of birds crash to the earth with the felling of the trees, "and the eggs within them were smashed, had fallen out, had mingled with the leaves and the dust, the little birds within the cracked eggs half-formed and dried up, dying as they were emerging into a hard, miraculous world" (242). Just as the eggs are smashed without the birds inside ever flying, the death of the trees robs the local community of sources for potential food, medicine, and wealth.

The transition from the forest as lived in and managed space into an impediment to the expansion of European-style urban space reveals one of the main tensions that Okri's trilogy explores. As deforestation alters the physical landscape of West Africa, the foodways that distinguish much of West African agricultural practice are torn from their moorings. The *Famished Road* trilogy traces the effects of the deforestation that accelerate as independence looms. In *The Famished Road*, the forest, while threatened, is still at the heart of the novel. Azaro follows paths into its depths, gets lost, and sees mythical creatures. By the second novel in the trilogy,

> trees were being felled every day in the forest. . . . The forest became dangerous. It became another country, a place of spectral heavings, sighs, susurrant arguments as of a council of spirit elders, a place with fleeting visions of silver elephants and white antelopes, a place where elusive lions coughed—a bazaar of the dead. And because the forest gradually became alien to us, because we feared the bristling potency of its new empty spaces, we all became a little twisted.[9]

The transition from the forest as lived-in space to alien space as it diminishes culminates in the hole left by its absence in *Infinite Riches,* where the centrality of the forest to the lives and myths of the region's inhabitants only becomes clear when the forest is completely cleared. It is easy to lose sight of the material reality underscoring the forest's destruction while absorbing Okri's dreamlike descriptions of forest spirits, singing white antelope, and lovely girls with bright green eyes. However, the links between deforestation and poverty are emphasized throughout the novels, not least when Azaro notes the pervasive lack of food: "I walked barefoot in a world breaking down under the force of hunger" (5–6).

That deforestation should be linked so directly to food production may come as somewhat of a surprise, because modern development narratives often place food production as a major cause of deforestation.[10] Traditionally, however, a great deal of agriculture in West Africa was agroforestry, in which cultivated crops were grown alongside a variety of tree species.[11] Today the great forest basin that swept across what is now southern Nigeria is nearly gone, with only 4.9 percent of land still covered by rain forest.[12] The ramifications of this deforestation are unknown; because the "'North' . . . is grain-rich [but] gene-poor,"[13] the potential for the kind of crop failure that devastated Ireland in the 1840s increases as tree and shrub varieties vanish. As an example, there are more than a thousand different species of oil palm (*Elaeis guineensis*) traditionally cultivated in West Africa, but only two species form the basis of the modern plantation system of oil production.[14] As an Akan proverb has it, the oil palm has innumerable purposes. Deforestation and the importation of plantation-style cultivation mean that local knowledge of the differences between cultivars—which species is better for which purpose—has been lost.

It is in this context that the depiction of the rain forest in the *Famished Road* trilogy takes on such significance. While some of the more unusual sights in Okri's forest come directly from Okri's own imagination, others have connections back to stories of the forest in various West African traditions. In her refutation of Douglas McCabe's reading of *The Famished Road* as New Age literature, Esther de Bruijn points out the connections that McCabe makes between the forest in Okri's novel and "the sinister 'bush' of the Ifa tradition and Amos Tutuola's *My Life in the Bush of Ghosts.*"[15] While others have experiences with the spirit world in the novel—Mum with the white man who turns black in the market, Dad with the man in the white suit—only Azaro can consistently see the spirits. While this second sight is inherent in him because of his identity as an *abiku,* he must learn to use it. In order to negotiate the spaces of the forest without becoming lost, Azaro must learn to see them. Only by looking with eyes informed by knowledge can he see the patterns of reproduction and decay in the forest. As Alfred Oteng-Yeboah points out in an essay on the philosophy of sustainable forest use among the Akan, traditional resource management in the forests of West Africa was a complex system involving traditional religious taboos relating to logging and farming practices as well as scientific knowledge, all "meant to ensure a sustainable harmony of human beings with their environment for the purpose of survival."[16] The loss of this way of seeing the forest results in the scene of destruction that Azaro views when he follows the sound of weeping from the forest and comes upon a group of loggers clearing the forest:

> Behind the men was a magnificent iroko tree. It was beautiful. Its trunk span was so vast that ten men couldn't link their arms around it. The iroko stood a third sawn through. The men had thick ropes attached to its higher parts.
> Men were shouting everywhere and the noise of weeping sounded all around like a giant in agony.[17]

Fleeing from the sobbing tree and the violence of the loggers, Azaro wanders deeper into the forest, where he discovers "a settlement, a cluster of white huts, with a fence all around them" (*Infinite,* 85). This settlement offers a glimpse into another way of living with the forest, one that does not involve clear-cutting millennia-old iroko trees. There is a danger here in too heavily romanticizing a precolonial way of living with the forest, as Kojo Sebastian Amanor reminds readers in his essay on Ghanaian farmers and forestry:

> Here, as elsewhere, there are no monolithic 'local communities' or undifferentiated 'forest societies.' At the village level, conflicting forestry-related viewpoints and agendas struggle constantly to gain hegemony, with dominant economic, political and institutional interests routinely claiming to represent either entire communities, or, more frequently, the 'common good.'[18]

As the *Famished Road* cycle insists, people need to eat, and often the only sources of sustenance involve activities that destroy their bodies, their souls, or their environment. While the timber from the fallen iroko is not going to benefit the local community, the loggers will be able to feed their families. Thus, any attempt at ecological intervention needs to take into account the necessity for labor and access to food. As Amanor concludes in his essay, "If 'development' is to meet the needs of rural people then environmental and sustainable 'development' concerns need to be integrated into people's livelihood strategies and production relations, and to be embedded in their political and economic struggles" (320). Azaro glimpses some of the potential for a new economic relation with the forest as he hallucinates under the branches of the giant iroko tree that has fallen under the loggers' saws, trapping him beneath the branches. The loggers have freed "forgotten diseases" from "calm quarantined contentment" (*Infinite*, 89). The cures to those forgotten diseases are only to be found in the forests from which they originated, and only the women of the forest, who "had discovered the secrets of herbs and bark" (*Songs*, 79), know those cures. Going into the forest with a sense of wonder, open to the forest's possibilities, serves as opposition to the global movement of capital, represented by the market where Azaro's father carries sacks of concrete on his back and by the subject of the title of the trilogy, the famished road itself.

As Azaro's father points out, deforestation will lead to institutionalized poverty. The road, along with its constant hunger for the movement of goods and bodies, represents the unequal balance of trade that comes to define African economies. Azaro's father links the hunger of the road to its origin as a river, with the demands, as it breaches its banks during the rainy season: "In the beginning there was a river. The river became a road and the road branched out to the whole world. And because the road was once a river it was always hungry" (*Famished*, 3).

Unlike portrayals of the road in African narratives that are linked to earlier portrayals of rivers, the famished road does not demand sacrifices because of its life-giving nature but merely because it is always hungry. Azaro's father tells the story of the King of the Road and how the people of the world band together to try to defeat that personification of hunger. The story is closely linked to the Swallowing Monster stories that exist across the African continent.[19] As in some of those stories the King of the Road is tricked into eating himself, but in Azaro's father's version "What had happened was that the King of the Road had become part of all the roads in the world. He is still hungry, and he will always be hungry" (*Famished*, 261). The famished road, then, is a representation of global capitalism. The road eats without discrimination, devouring whole communities in its mission to feed an insatiable external market.

The character of Azaro opens up both the spaces of the road and the forest for the novel's readers because he wanders through them, getting captured on the road by different people who want to sacrifice him for prosperity and getting lost in the forest among tree spirits and rampaging monsters. Azaro travels in other ways as well that serve to reveal the differences of the ideologies behind the representation of the road and the forest. In *Infinite Riches,* Azaro flits through what he terms "contending dreams," the hopes and plans and schemes that various constituencies have for the future of independent Nigeria or for all of Africa (201).

The continuation of the colonial system is represented by the British governor-general, who is determined that the devouring nature of the road continue for the benefit of Great Britain. To that effect, on the eve of Nigerian independence he rewrites African history. As he "reinvented the geography of the nation and the whole continent," everything that predates the arrival of the Europeans in Africa is effaced from history (110–11). The history that Nigerian children will receive even after independence will show them as inferior, with no "art, science, mathematics, sculpture, abstract conception, and philosophy" (111) before the Europeans bestowed these wonders on the continent. In the governor-general's dream, "a heroic and beautiful road" has been built under his supervision: "he dreamt that on this beautiful road all Africa's wealth, its gold and diamonds and diverse mineral resources, its food, its energies, its labours, its intelligence would be transported to his land to enrich the lives of his people across the green ocean" (204).

Opposing the effacement of history is the old woman of the forest. She represents "the true secret history" (112) of the peoples of the continent. In a private language "of signs and symbols, of angles and colours and forms, she recorded legends and moments of history lost to her people" (113). She dreams of decolonization and the rejection of imported culture and a turn instead to forms of government and ways of living suggested by the precolonial past. While her dreams offer hope for "an eventual, surprising, renaissance" (114) after the horrors of slavery, colonization, and neocolonial governments, the hope seems mediated by the fact that she records this history in a language that no one else can see and that her refuge in the forest is falling beneath loggers' axes. What good is the recording of the history and the uses of the forest if the forest itself vanishes?

While these two diametrically opposed visions of Africa can seem on the one hand stereotypical (there are echoes of Achebe's district commissioner from *Things Fall Apart* in Okri's character) and on the other hand overly romantic (the mysterious white antelopes and the forest hermit come closest to confirming Douglas McCabe's charge that the books' "most important cultural vector" is New Age spirituality),[20] the third character whose dreams Azaro visits absolutely resists easy categorization. If the governor-general represents a colonial vision and the old woman of the forest represents a precolonial vision, then Madame Koto serves as the embodiment of independence, with all the messiness and contradictions that this entails.

Visits into the dreams of Madame Koto frame the opposing dreams of the governor-general and the old woman of the forest. This seems appropriate for Madame Koto and her aspirations dominate the trilogy. It is her character who best captures the transition from forest to road and shows the allure of the global capital. Madame Koto owns one of one of the many chop joints on the outskirts of the city, and part of the appeal of her bar is her peppersoup, a poor person's dish made from peppers, greens, ginger and other spices, and meat that is served with yam. As Madame Koto's bar becomes the preferred hangout of politicians and their thugs, the food that it offers changes as well. The first indication of this is the Coca-Cola poster that appears on the wall of her bar with "the picture of a half-naked white woman with big breasts" (*Famished*, 215). Coca-Cola is not the only bottled beverage that Madame Koto sells: "Madame Koto graduated from palm wine to beer. There was more money in beer" (383). The peppersoup made with chicken heads is displaced by "long tables tumbled with fruit and fried meat, rice and platters of sweet-smelling stews, vegetables, and plastic cutlery" (455). The plastic cutlery is a clear indication of the increasing disparity between rich and poor over the course of the novel. "Poverty arrives with a plastic bowl," observes Renee Neblett, the founder of the Kokrobitey School in Ghana,[21] and it is no coincidence that beggars arrive in the novel at the same time that Madame Koto embraces both imported products and the Party of the Rich.

Despite her greed, Madame Koto is not a one-dimensional character. While she is aligned with the Party of the Rich and the cynical machinations of the poverty that many in the country feel on the eve of the first elections, she also is capable of great kindness toward those whom she sees as part of her community. In an essay titled "Madame Koto: Grotesque Creatrix or the Paradox of Psychic Health," Maggi Phillips links the portrayal of Madame Koto with "the same pattern of ambivalence which pervades the [Yoruba] female deity Oya."[22] Phillips quotes Judith Gleason on Oya's attributes, which include being simultaneously "On the side of death, on the side of life."[23] When the neighborhood begins to turn against her because of the violence of the thugs, she cries out in her own defense: "You all stare at me as if I am giving birth to a horse, but which one of you can give birth to a country and not die of exhaustion, eh?" (*Infinite*, 29). This explicit reference to her power being used for nation building explains the monstrous pregnancy that seems to have no end. Even after death her body keeps swelling until it has to be housed in "a coffin made of the hardest steel" (306). The postcolonial country that Madame Koto is trying to birth will not emerge. When Madame Koto cries out that she "cannot cut down old trees" because "they give shade to two thousand caravans of spirits" (30), she ignores the fact that the colonial forces have been particularly efficient at deforestation and that the iroko as well as the acacia fall under the colonizers' axes. As with the forest, whose worth is only comprehended after its loss, Madame Koto's significance in the community is only understood after she is killed: "There are certain trees that seem worthless but when gone leave empty spaces through which bad winds blow. There are other trees that seem useless but when felled worse things grow in their place" (283). Clearly in light of the context of the coming election, Madame Koto's death prefigures a cycle of worsening governments that offer hope to the dispossessed just to snatch it from them again. "An abiku nation, a spirit-child nation, one that keeps being reborn and after each birth come blood and betrayals" (*Famished*, 494). And as this bloody cycle continues, the stature of leaders is necessarily diminished to the advantage of those at the other end of the famished road.

After Madame Koto's death, the original bar girls she had saved from violence and depravity return to prepare her body for burial. All have become successful because Madame Koto "had opened up their roads for them" (*Infinite*, 312). While they may have had to prostitute themselves for Madame Koto in order to build a new life, the shade of her presence allowed them to flourish: "And they prepared the feast, cleaned Madame Koto's rooms, organized the orderly dismantling of her realm, and determined for her an honourable funeral, because they knew that great old trees are impossible to replace" (312). While her generosity to these women has freed them from tyranny, Okri seems to argue that it is a misuse or at least a misdirection of the potential that Madame Koto holds. When Azaro enters her dream, he sees it as too narrow and as only one voice in a river of dreams: "The dreams were too many, too different, too contradictory: the nation was composed not of one people but of several mapped and bound into one artificial entity by Empire builders" (202). The road may have helped Madame Koto's bar girls escape bad situations, but Madame Koto's allegiance to the road and the forces of globalization rob the community of control over local resources, thus creating a situation that leads to more poverty and increasing stress for and violence against women.

The reference to dreams in the third book of the trilogy invokes the closing of *The Famished Road,* which ends with the line "A dream can be the highest point of life" (500). That closing references both Dad's dream of redreaming the world and the potential of a new standard of living with the arrival of independence. In his dreams, Azaro's father

> saw the world in which black people always suffered and he didn't like it. He saw a world in which human beings suffered so need-lessly from Antipodes to Equator, and he didn't like it either. He saw our people drowning in poverty, in famine, drought, in divi-siveness and the blood of war. He saw our people always preyed upon by other powers. Manipulated by the Western world, our history and achievements rigged out of existence. (492)

While Dad lies in a coma pleading for justice in all the courts of the universe, Azaro's mother "prayed for simple things that made me weep while the darkness flowered in our room. She prayed for food. She prayed for Dad to get well. She prayed for a good place to live. She prayed for more life and for suffering to bear lovely fruits. And she prayed for me. For three days Mum prayed on borrowed wine" (493). Dad's dreams and Mum's prayers are directly related; the injustices that Dad rails against in the courts of the universe and the basic necessities that Mum prays for emerge from the same economic system that displaces local control over food and other material necessities in favor of the removal of resources from the local area to benefit a distant elite. Dad's dream makes it clear that while the system of exploit-ative trade was created by "white people," it will be continued by the rich and the politicians of independent Africa. The problem with both Dad's dreams and Mum's prayers are that they take place alone. Like Madame Ko-to's dream of giving birth to a nation, they are all lost in the "feverish con-fluence of contending waters" (*Infinite*, 202). The isolation only benefits the continued dispossessive organization of the world: "Dad was alone because he didn't see the others, the multitudes of dream-pleaders, invading all the courts of the universe, while struggling in the real hard world created by the limitations in the minds of human beings" (*Famished*, 493). Okri's novels suggest that for a dream not to be the highest point of life, those multitudes must come together in that "real hard world," with imaginations freed from any limitations. Or, as he writes in *Songs of Enchantment*, "A dream can be the highest point of life; action can be its manifestation" (275).

That connection between dreams and action can be seen when Azaro frees a duiker at Madame Koto's feast. Throughout the novel, food prod-ucts originating from the forest have a positive connotation, from the wild boar that Azaro's father kills to celebrate his son's return to the dongayoro used to treat illness and the palm wine that keeps despair at arm's length. At the feast to celebrate Madame Koto's "consolidation of her party connections" (*Famished*, 449), a duiker is tied to a post, waiting to be sacrificed, although there is more than enough meat on the table to both feed the crowd and appease the gods. Azaro stares into the duiker's eyes and experiences the history of man through nonhuman consciousness. Hunting and being hunted have always defined the relationship of humans and duikers, but once "the beings of an earlier time were creators first before they were hunters" (456). This changes with the arrival of "the ghost ships of centuries," which bring enslavement and a new way of seeing the relationship between hunter and hunted, a new divide between culture and nature: "The white ones, ghost forms on deep nights, stepped on our shores, and I heard the earth cry" (457). With the arrival of the white ones, as his mother has indicated to him in another story, the relationship between humans and the earth changed, with both the humans and the forests diminished. The sacrifice of the duiker tied to a pole represents another diminishment. "When human beings and animals understood another, we were all free," Azaro thinks (457). His subsequent release of the antelope has to be read as a symbolic attempt to free both the human and nonhuman from the greed of global capitalism. It is the beginning of the revolution that Dad hopes to sow "like a fertile seed in the earth" (*Infinite*, 200).

While that revolution might seem to be a pipe dream, there are groups trying to restore local economies, focusing particularly on the systems of foodways and localized ecologies. Recognizing perhaps that "The only power poor people have is their hunger" (*Famished*, 70), groups are coming together across the former colonized regions with agendas focusing on food sovereignty. The international peasants' rights organization La Via Campesina has articulated the principle of food sovereignty as

> "The Right of the Peoples to produce their own foods and organize food production and consumption according to the needs of local communities placing priority on the production and consumption of local domestic products." . . . [This includes] the right of every man and women to have access to resources to produce their own foods, maintain their productive culture and preserve their food culture and above all their national sovereignty.[24]

As studies have shown, the loss of control over local crop selection and foodways leads to malnutrition in the very people producing food, often because the type of food grown has altered or because more is being produced as a commodity to be sold to wealthier markets.[25] Organizations such as Le Réseau des organisations paysannes et de producteurs de l'Afrique de l'Ouest (ROPPA) advocate interventions that raise the profile of regional foods across West Africa, providing farmers incentives to grow local foods and advocating the substitution of West African crops for imported grains and tubers in foodstuffs produced for the local market.[26] Smallholder farmers' associations advocate diversification of farming away from monocultural crops toward a wider variety of crops, clearly articulating the connections between food production and consumption and gender relations and disease prevention while working exclusively with smallholder farmers who are likely to face "persistent chronic food insecurity."[27] Food sovereignty, then, is one of the defining principles of true liberation in a Fanonian sense, and food sovereignty in West Africa is inextricably bound up with forest use and protection because of the long history of agroforestry across the region. In forest regions communities have long and deep relationships to the forest that predate the colonial and neocolonial eras. Those relationships vary widely, but most local communities across West Africa lost significant control over forest resource use in the late colonial period, a situation that continued after independence.[28] In his essay in this volume, Rob Nixon uses the wonderfully evocative phrase "slow violence" to describe the impact that globalization had and continues to have on the rural populations of Africa. The trucks laden with logs that daily depart from the Congo Basin for seaports across West Africa represent more than just the alteration of the landscape through deforestation.

As Melissa Leach points out, "In losing control over the capacity to decide their use of forest resources, people therefore lose control over their lives in much more fundamental ways" (227). The impact of that deforestation on the communities of West Africa—the slow violence the Nixon excoriates—can be seen in nutrient-deficiency diseases such as kwashiorkor, marasmus, and pellagra and in the overwhelming dominance of the informal economic sector across the region.[29] This impact can also be seen in the concentration of wealth in the hands of a few people in the community as what was once common land is increasingly privatized.[30]

Nixon turns to the work of the Green Belt Movement in Kenya to show how the deceptively simple act of tree planting engaged economic and social forces to such an extent that planting a tree became "an incendiary, seditious act of civil disobedience" (see Nixon, this volume). In her Nobel Peace Prize lecture, Wangari Maathai, the founder of the Green Belt

Movement, explained how an advocacy group for rural women and reforestation became involved in issues of democracy and fair governance:

> Although initially the Green Belt Movement's tree planting activities did not address issues of democracy and peace, it soon became clear that responsible governance of the environment was impossible without democratic space. Therefore, the tree became a symbol for the democratic struggle in Kenya. Citizens were mobilised to challenge widespread abuses of power, corruption and environmental mismanagement.[31]

The mobilization offers hope for what Frantz Fanon termed the true liberation of the colonized because men and women control "all the material means which make possible the radical transformation of society" (310). The relations of humans with matter, with the world outside, and with history are still in the neocolonial period simply relations with food. While redreaming the world and freeing the duiker to wreak havoc among the forces of globalization are important first steps, they are only prayers offered with borrowed wine. The local cycles of production and consumption need to be restored, forests need to be replanted, and crops need to be remembered so that food comes under the control of those who plant it for their tables, for their families and their friends, and for their livelihood and their very survival.

Notes

Thanks to Anthony Vital, Byron Caminero-Santangelo, Rob Nixon, and Nicole Merola for their thoughtful comments on earlier drafts of this essay.

1. Frantz Fanon, *The Wretched of the Earth* (New York: Grove, 1968), 308.

2. Ibid., 152, quoted in Khabisi Mosunkutu, "Gauteng Agriculture, Conservation and Environment Prov Budget Vote," June 6, 2006, http://www.polity.org.za/article.php?a_id=87697 (accessed May 3, 2010).

3. Mosunkutu, "Gauteng."

4. René LeMarchand, "The Political Economy of Food Issues," in *Food in Sub-Saharan Africa,* edited by Art Hansen and Della E. McMillan (Boulder, CO: Lynne Rienner, 1986), 27.

5. James C. McCann, *Maize and Grace: Africa's Encounter with a New World Crop, 1500–2000* (Cambridge: Harvard University Press, 2005), 7.

6. Hope Shand, *Human Nature: Agricultural Biodiversity and Farm-Based Food Security* (Ottawa: Rural Advancement Foundation International, 1997), 2.

7. Ben Okri, *The Famished Road* (New York: Doubleday, 1991), 408.

8. In *Songs of Echantment*, Azaro's father tells his family of the explosion of "a big bomb in our backyard" (146). and a subsequent heat wave, a reference to the first French nuclear test, Gerboise Bleu, detonated in Algeria on February 13, 1960. See http://cr4.globalspec.com/blogentry/1205/February-13-1960-France-Tests-an-A-Bomb (accessed May 3, 2010).

9. Ben Okri, *Songs of Enchantment* (New York: Doubleday, 1993), 68.

10. For a discussion of the pervasive idea in development studies that African farmers are a primary cause of deforestation and desertification, see Vigdis Broch-Due, "Producing Nature and Poverty in Africa: An Introduction," in *Producing Nature and Poverty in Africa*, edited by Vigdis Broch-Due and Richard A. Schoeder, 9–52 (Stockholm: Nordiska Afrikainstitutet, 2000).

11. John A. Poku, "Management of Trees in Association with Crops in Traditional Agroforesty Systems," in *Managing Agrodiversity the Traditional Way: Lessons from West Africa in Sustainable Use of Biodiversity and Related Natural Resources*, edited by Edwin A. Gyasi, Gordana Kranjac-Berisavljevic, Essie T. Blay, and William Oduro (New York: United Nations University Press, 2004), 155.

12. "Nigeria May Be Left without Forest by 2010," *Terra Daily: News About Planet Earth*, January 18, 2007, http://www.terradaily.com/reports/Nigeria_May_Be_Left_Without_Forest_By_2010_999.html (accessed May 3, 2010).

13. Cary Fowler and Pat Mooney, *Shattering: Food, Politics, and the Loss of Genetic Diversity* (Tucson: University of Arizona Press, 1990), xi.

14. Lawrence K. Opeke, *Tropical Tree Crops* (Chichester, UK: Wiley, 1982), 252. See also A. Hayati, R. Wickneswari, I. Maizura, and N. Rajanaidu, "Genetic Diversity of Oil Palm (*Elaeis guineensis* Jacq.) Germplasm Collections from Africa: Implications for Improvement and Conservation of Genetic Resources," *TAG: Theoretical and Applied Genetics* 108, no. 97 (May 2004): 1274.

15. Esther de Bruijn, "Coming to Terms with New Ageist Contamination: Cosmopolitanism in Ben Okri's *The Famished Road*," *Research in African Literatures* 38, no. 4 (2007): 174. See also Douglas McCabe, "'Higher Realities': New Age Spirituality in Ben Okri's *The Famished Road*," *Research in African Literatures* 36, no. 4 (2005): 1–21, and Douglas McCabe, "Forum: Douglas McCabe's Response to Esther de Bruijn's Essay," *Research in African Literatures* 38, no. 4. (2007): 227–33.

16. Alfred A. Oteng-Yeboah, "Philosophical Foundations of Biophysical Resource Use with Special Reference to Ghana," in *Managing Agrodiversity the Traditional Way*, 8.

17. Ben Okri, *Infinite Riches* (London: Phoenix House, 1998), 84.

18. Kojo Sebastian Amanor, "Farmers, Forestry and Fractured Environmentalisms in Ghana's Forest Zones," in *Contesting Forestry in West Africa*, edited by Reginald Cline-Cole and Clare Madge (Aldershot, UK: Ashgate, 2000), 307.

19. Alice Werner, *Myths and Legends of the Bantu* (London: Frank Cass, 1968) 163.

20. McCabe, "Higher Realities," 2.

21. Renee Neblett, personal conversation, October 2007.

22. Maggi Phillips, "Madame Koto: Grotesque Creatrix or the Paradox of Psychic Health," in *Seriously Weird: Papers on the Grotesque*, edited by Alice Mills (New York: Peter Lang, 1999), 40.

23. Judith Illsley Gleason, *Oya: In Praise of the Goddess* (Boston: Shambhala, 1987), 51, quoted in Phillips, "Madame Koto," 41.

24. Ramiro Tellez, "Why Food Sovereignty in the Bolivian Constitution?," *La Via Campensina*, May 22, 2007, http://www.viacampesina.org/main_en/index.php?option=com_content&task=view&id=316&Itemid=38 (accessed January 22, 2008).

25. David Barkin, Rosemary L. Batt, and Billie R. DeWalt, *Food Crops vs. Feed Crops: Global Substitution of Grains in Production* (Boulder, CO: Lynne Rienner, 1990), 1.

26. Jacques Berthelot, "Food Sovereignty, Agricultural Prices, and World Markets," paper presented at the Forum on Food Sovereignty held in Niamey, Niger, November 7–10, 2006; reprinted at http://www.roppa.info/IMG/pdf/J._Berthelot-Food_sovereignty_agricultural_prices_and_world_markets-ROPPA_November_06.pdf (accessed July 3, 2010).

27. Sam Moyo, quoted in Nosimilo Ndlovu, "Growing Our Own Food," *Mail and Guardian online*, August 26, 2008, http://www.mg.co.za/article/2008-08-26-growing-our-own-food (accessed May 3, 2010). See also National Smallholder Farmers' Association of Malawi, http://www.nasfam.org/index.php?option=com_frontpage&Itemid=1 (accessed May 3, 2010).

28. See Reginald Cline-Cole, "Redefining Forestry Space and Threatening Livelihoods in Colonial Northern Nigeria," in *Contesting Forestry in West Africa*, 36–63; James Fairhead and Melissa Leach, "Shaping Socio-Ecological and Historical Knowledge of Deforestation in Sierra Leone, Liberia and Togo," in *Contesting Forestry in West Africa*, 64–95; Melissa Leach, *Rainforest Relations* (Washington, DC: Smithsonian Institute Press, 1994), 8–11; and Tony Binns, *Tropical Africa* (London: Routledge, 1994), 65–67.

29. "Fact Sheets: The Informal Economy," WIEGO, Women in Informal Employment: Globalizing and Organizing, http://www.wiego.org/main/fact1.shtml (accessed May 3, 2010).

30. Kathleen M. Baker, *Indigenous Land Management in West Africa: An Environmental Balancing Act* (Oxford: Oxford University Press, 2000), 63.

31. Wangari Maathai, "Nobel Lecture," Oslo, December 10, 2004, http://nobelprize.org/nobel_prizes/peace/laureates/2004/maathai-lecture-text.html (accessed January 22, 2008).

Bibliography

Amanor, Kojo Sebastian. "Farmers, Forestry and Fractured Environmentalisms in Ghana's Forest Zones." In *Contesting Forestry in West Africa*, edited by Reginald Cline-Cole and Clare Madge, 307–21. Aldershot, UK: Ashgate, 2000.

Baker, Kathleen M. *Indigenous Land Management in West Africa: An Environmental Balancing Act.* Oxford: Oxford University Press, 2000.

Barkin, David, Rosemary L. Batt, and Billie R. DeWalt. *Food Crops vs. Feed Crops: Global Substitution of Grains in Production.* Boulder, CO: Lynne Rienner, 1990.

Berthelot, Jacques. "Food Soveignty, Agricultural Prices, and World Markets." Paper presented at the Forum on Food Sovereignty held in Niamey, Niger, November 7–10, 2006. Reprinted at http://www.roppa.info/IMG/pdf/J._Berthelot-Food_sovereignty_agricultural_prices_and_world_markets-ROPPA_November_06.pdf.

Binns, Tony. *Tropical Africa.* London: Routledge, 1994.

Broch-Due, Vigdis. "Producing Nature and Poverty in Africa: An Introduction." In *Producing Nature and Poverty in Africa*, edited by Vigdis Broch-Due and Richard A. Schoeder, 9–52. Stockholm: Nordiska Afrikainstitutet, 2000.

Cline-Cole, Reginald. "Redefining Forestry Space and Threatening Livelihoods in Colonial Northern Nigeria." In *Contesting Forestry in West Africa*, edited by Reginald Cline-Cole and Clare Madge, 36–63. Aldershot, UK: Ashgate, 2000.

de Bruijn, Esther. "Coming to Terms with New Ageist Contamination: Cosmopolitanism in Ben Okri's *The Famished Road.*" *Research in African Literatures* 38, no. 4 (2007): 170–86.

"Fact Sheets: The Informal Economy." WIEGO (Women in Informal Employment: Globalizing and Organizing), http://www.wiego.org/main/fact1.shtml.

Fairhead, James, and Melissa Leach. "Shaping Socio-Ecological and Historical Knowledge of Deforestation in Sierra Leone, Liberia and Togo."

In *Contesting Forestry in West Africa,* edited by Reginald Cline-Cole and Clare Madge, 64–95. Aldershot, UK: Ashgate, 2000.

Fanon, Frantz. *The Wretched of the Earth.* New York: Grove, 1968.

Fowler, Cary and Pat Mooney. *Shattering: Food, Politics, and the Loss of Genetic Diversity.* Tucson: University of Arizona Press, 1990.

Hayati, A., R. Wickneswari, I. Maizura, and N. Rajanaidu. "Genetic Diversity of Oil Palm (*Elaeis guineensis* Jacq.) Germplasm Collections from Africa: Implications for Improvement and Conservation of Genetic Resources." *TAG: Theoretical and Applied Genetics* 108, no. 97 (May 2004): 1274–84.

Illsley Gleason, Judith. *Oya: In Praise of the Goddess.* Boston: Shambhala, 1987.

Leach, Melissa. *Rainforest Relations.* Washington, DC: Smithsonian Institute Press, 1994.

LeMarchand, René. "The Political Economy of Food Issues." In *Food in Sub-Saharan Africa,* edited by Art Hansen and Della E. McMillan, 25–43. Boulder, CO: Lynne Rienner, 1986.

Maathai, Wangari. "Nobel Lecture." Oslo, December 10, 2004, http://nobelprize.org/nobel_prizes/peace/laureates/2004/maathai-lecture-text.htm.

McCabe, Douglas. "Forum: Douglas McCabe's Response to Esther de Bruijn's Essay." *Research in African Literatures* 38, no. 4 (2007): 227–33.

———. "'Higher Realities': New Age Spirituality in Ben Okri's *The Famished Road.*" *Research in African Literature* 36, no. 4 (2005): 1–21.

McCann, James C. *Maize and Grace: Africa's Encounter with a New World Crop, 1500–2000.* Cambridge: Harvard University Press, 2005.

Mosunkutu, Khabisi. "Gauteng Agriculture, Conservation and Environment Prov Budget Vote," June 6, 2006, http://www.polity.org.za/article.php?a_id=87697.

National Smallholder Farmers' Association of Malawi, http://www.nasfam.org/index.php?option=com_frontpage&Itemid=1.

Ndlovu, Nosimilo. "Growing Our Own Food." *Mail and Guardian online,* August 26, 2008, http://www.mg.co.za/article/2008-08-26-growing-our-own-food.

"Nigeria May Be Left without Forest By 2010." *Terra Daily: News About Planet Earth,* January 18, 2007, http://www.terradaily.com/reports/Nigeria_May_Be_Left_Without_Forest_By_2010_999.html.

Nixon, Rob. "Slow Violence, Gender, and the Environmentalism of the Poor," in *Environment at the Margins: Literary and Environmental Studies in Africa,* edited by Byron Caminero-Santangelo and Garth Myers. Athens: Ohio University Press, 2011.

Okri, Ben. *The Famished Road.* New York: Doubleday, 1991.

———. *Infinite Riches.* London: Phoenix House, 1998.

———. *Songs of Enchantment.* New York: Doubleday, 1993.

Opeke, Lawrence K. *Tropical Tree Crops.* Chichester, UK: Wiley, 1982.

Oteng-Yeboah, Alfred A. "Philosophical Foundations of Biophysical Resource Use with Special Reference to Ghana." In *Managing Agrodiversity the Traditional Way: Lessons from West Africa in Sustainable Use of Biodiversity and Related Natural Resources,* edited by Edwin A. Gyasi, Gordana Kranjac-Berisavljevic, Essie T. Blay, and William Oduro, 8–13. New York: United Nations University Press, 2004.

Phillips, Maggi. "Madame Koto: Grotesque Creatrix or the Paradox of Psychic Health." In *Seriously Weird: Papers on the Grotesque,* edited by Alice Mills, 35–49. New York: Peter Lang, 1999.

Poku, John A. "Management of Trees in Association with Crops in Traditional Agroforesty Systems." In *Managing Agrodiversity the Traditional Way: Lessons from West Africa in Sustainable Use of Biodiversity and Related Natural Resources,* edited by Edwin A. Gyasi, Gordana Kranjac-Berisavljevic, Essie T. Blay, and William Oduro, 155–64. New York: United Nations University Press, 2004.

Shand, Hope. *Human Nature: Agricultural Biodiversity and Farm-Based Food Security.* Ottawa: Rural Advancement Foundation International, 1997.

Tellez, Ramiro. "Why Food Sovereignty in the Bolivian Constitution?" *La Via Campensina,* May 22, 2007, http://www.viacampesina.org/main_en/index.php?option=com_content&task=view&id=316&Itemid=38.

Werner, Alice. *Myths and Legends of the Bantu.* London: Frank Cass, 1968.

Chapter 7

Whites Lost and Found

Immigration and Imagination in Savanna Africa

David McDermott Hughes

> I had a farm in Africa. . . . The views were immensely wide. Everything
> that you saw made for greatness and freedom, and unequalled
> nobility. . . . In the Highlands you woke up in the morning and thought:
> Here I am, where I ought to be.
>
> **Isak Dinesen,** *Out of Africa*

> I have sometimes thought since of the Elkingtons' tea table—round,
> capacious, and white, standing with sturdy legs against the green vines
> of the garden, a thousand miles of Africa receding from its edge.
> It was a mark of sanity.
>
> **Beryl Markham,** *West with the Night*

SYMPATHETIC AUTHORS frequently describe African whites as a "lost
tribe."[1] The phrase suggests a population marooned, wandering, or scattered
(from Israel) or otherwise out of step with its surroundings. Indeed, in this
metaphorical sense, Europeans partly failed as settlers and immigrants to
Africa in the twentieth century. Of course, at various times and places they
nearly monopolized power, wealth, and/or land. But in the realm of ideas, few
could convince themselves and others that they belonged. Barronness Blixen,
under the pen name Isak Dinesen, wrote with unmatched certainty when
she declared, "Here I am, where I ought to be." (If everyone knew it be true,
of course, she would not have needed to say it, and bankruptcy sent her back
to Denmark anyway.) At almost the same time and about the same Kenyan

landscape, Beryl Markham gave voice to a deeper ambivalence and fear: beyond the veranda of colonial control lay a strange, uncontrolled vastness. To understand and depict that world, whites engaged in what I call the "imaginative project of colonialism."[2] Particularly the writers among them imagined European immigrants living, working, and becoming one with African savanna. From roughly 1930 onward, this uncoordinated, unplanned literary effort gave shape to an ethic and sensibility of landscape. As Anthony Vital and Byron Caminero-Santangelo both suggest in this volume, such artistic work runs tangent to explicit politics. Yes, a sense of belonging encouraged white settlers to stay in Africa and to dominate Africans. "Literature," as Edward Said argues, "participat[es] in Europe's overseas expansion and . . . creates . . . 'structures of feeling' that support, elaborate, and consolidate the practice of empire."[3] But literature and the imagination are not reducible to a struggle for or against power. Settler fiction, memoir, and travelogue ran on a separate track, sometimes dangerously inattentive to nationalism and other political currents. Writers undertook an endeavor that, to their white readers, was more emotive and profound: to *find* whites in Africa. In ways that were partial, fleeting, and only semiconscious, white Africans convinced themselves and others that they belonged on this savanna.

Nowhere was this process of literary integration more necessary than in British East Africa and southern Africa. The settler colonies of Kenya and Southern Rhodesia (now Zimbabwe) suffered from a mismatch of power and population, what Dane Kennedy calls a "demographic conjuncture."[4] Having pacified native polities by 1900, whites sought to monopolize politics and the economy as settlers had done in the United States and Australia. Yet they immigrated in numbers far smaller than on those frontiers. White enclaves never topped 1 percent of the national population in Kenya, 5 percent in Zimbabwe, and 20 percent in modern South Africa.[5] Minority status led to fear and restraint in cultural expression. Whereas Americans of the eastern seaboard began to romanticize Indians in the nineteenth century, when most Indians were safely exterminated or expelled, white Africans preferred not to dwell on the native masses surrounding them.[6] And they chose not to dwell *with* them either. In contrast to French, Portuguese, or Dutch administrators, British colonial officers sought to prohibit rather than shape social and sexual intercourse.[7] Such regulations kept intermarriage and even the learning of African languages to a minimum.[8] In this context of self-imposed isolation, writers from the late 1930s onward implicitly took responsibility for adapting Euro-Africans to Africa.[9] And these authors did so on broadly environmental terms. Female writers frequently romanticized the savanna and its wildlife, expressing love, yearning, and rejection. With many exceptions, a smaller number of male authors conveyed a similar man-land bond through narratives of exploration and adventure on vivid African topographies. By writing landscape in these and other ways, writers and their readers overcame the feeling of territorial exile. Also, by fixating on the land, settlers put out of their minds the social exile in which they lived. Colonial literature, then, promoted a *selective* assimilation to Africa.

African nationalism could easily have scrambled the neat distinctions underpinning this sense of comfort but, in fact, did the opposite. Whites continued to write and read as if land forms mattered more than social forms. Undoubtedly, independence movements in the 1950s thrust black majorities into political prominence and, very quickly, into power. Settler populations ceded from north to south: Kenya in 1963, Zambia in 1965, Zimbabwe in 1980, and South Africa in 1994. The new black governments confronted whites and their privileges as contradictions to be resolved with varying degrees of tolerance and force. Little compulsion was required vis-à-vis the gentleman farmers of Kenya's "white highlands." Derided by Kenya's leading (black) writer as "parasites in a paradise," these wealthy settlers repatriated themselves in the 1960s and 1970s with comparative ease.[10] The less endowed, harder working, and more invested Euro-Zimbabwean farmers mounted stiffer resistance. Whites clung to their parcels through the incomplete land reform of the 1980s. In 2000, a paramilitary program of farm occupations sent most of them off the land to the main cities or abroad. Meanwhile, since the advent of black rule in 1994, South African whites have also been slowly emigrating. In short, power and population have come into line, and whites can hardly ignore their own minority status. Yet to a surprising degree, the writers among them have continued to marginalize blacks in their texts. As before, authors and their characters find solace in nature, but now nature represents something more: a force of greater moral good and historical transcendence than states, land reform, and the like. In more practical terms, whites still lead most regional conservation nongovernmental organization, dominate the burgeoning ecotourism business, and publish texts and photos related to these activities. By writing and in writing, then, postcolonial whites have mastered the ways in which "race and nature work as a terrain of power."[11] For them, nature naturalizes better than empire ever did.

Escaping African People

For European overseas settlers, bilateral human-land relationships frequently emerged from more complex triangular systems. In a comparison of the United States and South Africa, for instance, George Fredrickson emphasizes a process with three terms: *colonizers* "struggle with the *original occupants* for possession of the *land.*"[12] Before genocide emasculated the original occupants, Euro-Americans *did* engage with them. In the Great Lakes region, seventeenth-century settlers and Indians established a "middle ground" of shared politics, kin networks, and even religion.[13] Surely white Africans could have done the same. The earliest—or perhaps protoimperial—settlers did intermarry. The Portuguese *prazeiros* of sixteenth-century Mozambique ascended to local chieftainships, eventually losing all European ties.[14] Much later, colonial governments imposed a strict racial order, segregating blacks into rural reserves and urban townships and reducing intercultural contact to a minimum. However, colonial states could not segregate history and meanings. In many rural areas amid and around white settlement, memories of natives—and often natives themselves—litter the landscape. Upon entering the native reserve, the girl in a Doris Lessing story experiences "meaningless terror" and senses "a queer hostility in the landscape . . . [that] seemed to say to me: you walk here as a destroyer" of African society.[15] Africans, as the third point of the triangular relationship, would not go away. With effort, though, writers more loyal to the project of settlement than Lessing could minimize and ignore them. Thus, the bulk of Euro-African canon escapes African people while embracing African land.

It was not always so: blacks initially impressed themselves violently upon white bodies and imaginations. Around the turn of the twentieth century, blacks "rose" against settlers and their institutions. Most notably, Southern Rhodesia's 1896–97 Chimurenga killed 10 percent of the settler population.[16] Even when pacified as a group, black individuals still generated a sense of threat. The small white population depended on blacks for all manual labor. Blacks therefore circulated in all white spaces, including the home when the man of the house was away. How could settler society protect itself from the "black peril" of an African man raping a European white?[17] How could the same society police itself against the transgression of poor whites, lonely men, or the morally infirm who might wish to share the bed of a black person? In the first decades of the twentieth century, both Rhodesia and Kenya lurched between "outbursts of public hysteria." If a white woman was involved, black crime ranging from burglary to "insults" could easily generate a charge of attempted rape (Kennedy, *Islands,* 141–46). With greater clarity, Daphne Rooke's novel *Mittee* narrates the rape of a white woman and the prompt and righteous lynching of her assailant.[18] By midcentury, however, the increasing density of settlers had reduced white women's apparent vulnerability, and readers were ready to move on. Lessing's *The Grass Is Singing* (1950) complicates the black peril: a manservant hardly notices as the farm wife, going mad, undresses before him.[19] Later, without sexual motive, he kills her. Thereafter, rape virtually exited the scene of white fantasy. The subject only reappeared in literature when Coetzee's *Disgrace*—in which black men gang-rape a white woman—jolted the South African public in 1999.[20] For nearly five decades, then, white writers forewent the most lurid fantasies of multiracial life. They also forewent the more prosaic realities. "We actually see blacks differently," confessed Gordimer in 1983, "which includes, [due to racial segregation] not seeing, not noticing their unnatural absence.."[21]

Empty land discourse helped entrench this social blindness. Notions of *terra nullius* had, of course, accompanied and justified Europe's global expansion for centuries.[22] In New England, for example, seventeenth-century Puritans considered the landscape to be utterly wild. Indians lived there, but they were not cultivating fixed, fenced parcels. In other words, they did not "occupy" the land, and colonists felt unconstrained by their presence.[23] Later chroniclers often erased that presence altogether, transferring a myth of empty land to African frontiers (Fredrickson, 35). This precedent of the "New World" surfaces almost imperceptibly in the reminiscences of Elspeth Huxley, the Kenyan preeminent settler-writer. "Each time you came to [a glade]," she recalls, "you had the feeling that you were the first person to stand upon that verge and gaze across the tufted grasses, like Cortez and the Pacific, and that some extraordinary prehistoric animal would be browsing there."[24] She forgets both the Aztecs massacred by Cortez and the Kikuyu who surround her own farm, emphasizing instead a notion of Pleistocene biodiversity.[25] Indeed, Africa came to epitomize what Anne McClintock calls "anachronistic space," that is, a zone stuck "in a permanently anterior time."[26] Writing in 1942, for instance, Kenyan settler Beryl Markham finds the hunting grounds east of Nairobi "formless": "It was the way the firmament must have been when the waters had gone . . . It was an empty world because no man had yet joined sticks to make a house or scratched the earth to make a road or embedded the transient symbols of his artifice in the clean horizon" (236). This ideal of an unblemished void, in other words, suggested the persistence of biblical or otherwise primitive states and swept the clutter of Africans—actually traversing the savanna with cattle and goods—out of whites' imagination.

The void could even obscure such abominations as black nationalism, black rule, and friendship with blacks. After World War II, writers responded to political agitation with irony and indirection. Appearing in 1959, Huxley's

childhood memoir *Flame Trees of Thika* foreshadows the Mau Mau Revolt
of 1952–60 in only the most oblique fashion. Just arrived from England in
1914, Lettice remarks on whites' overconfidence: "why people should be so
much more nervous about wild animals . . . than about other human beings, I've never been able to understand" (67). Only long after the event, in
the third book of the trilogy begun with *Flame Trees*, did Huxley address
the reality of Mau Mau and the fantasy of antiwhite reprisals: "the night
of the long knives" (*Out*, 203). By default, then, the job of describing Mau
Mau and black rule—as it emerged—fell to an American, Robert Ruark.[27]
His novels *Something of Value* and *Uhuru* focus unremittingly on terror
inflicted by blacks. In this context, Ruark almost criticizes whites' passion
for nature. An American tourist declares that "If Kenya white people had
devoted a fraction of the time and effort you spend trying to convince a
lousy leopard to come to a tree so you can shoot him . . . [to instead] trying to make *people* out of savages . . . you wouldn't have all these troubles
in Africa today."[28] After the "troubles" culminating in independence and a
capitalist state, many white Kenyans reached across the color bar. Yet to the
extent that it exists, black-white friendship has hardly disrupted the escap-
ist narrative. Perhaps it requires too much emotional and representational
work, as compared with the land. "Untouched landscapes are undemand-
ing," writes Kuki Gallman, who moved from Italy to a ranch in northern
Kenya in 1970, eventually becoming a noted conservationist. When her son
died, "Nothing was expected of me by the ancient silence of the mountains
and of the mysterious gorges. In their unjudging, harmonious existence
I found again my own identity and my place."[29] Grieving, Gallman seeks
comfort not among Kenyans but in an empty land.

No novel explains—and criticizes—this evasion more explicitly than
Francesca Marciano's recent work, *Rules of the Wild*.[30] Loosely autobiographical, the narrative follows an Italian immigrant, Esmé, as she enters and as-
similates to the white, largely conservationist community of Nairobi. First,
the land startles her. "You are flattened between the immensity of the air
above you and the solid ground. It's all around you, 360 degrees" (15). Esmé
falls for Adam, a safari guide of settler stock. Later in the same district she
marvels at Adam's sense of belonging, an entitlement that he shares with
another Kenya-born white, Iris. "The[ir] conversation made me think of the
Aborigines in Australia who believe that the earth was shaped because it had
been sung. . . . That night, Adam and Iris sang their Northern Territory song-
lines, as if they had both owned it by birthright" (93). Can whites become
indigenous? Adam, at one with the land, suggests such a possibility, but Esmé
soon sees through him and his kind. "For the majority of people, *whites*, I

mean, the whole point of living in this country is to avoid the sight of other human beings.... If we could press a button and pulverize the humans who happen to spoil the view, we'd happily press it. That's the whole point of going on safari, isn't it?" (194, emphasis in original). Esmé replaces Adam with a journalist, social critic, and witness of the Rwandan genocide, ultimately losing many of her friends. Marciano herself appears to have upset a good proportion of Nairobi whites who recognized themselves under pseudonyms.[31] Many perhaps also resented her for raising precisely the question usually dodged: "How does a white person live in a black country? ... Where do we stand in relation to each other?" (100). For her, at least, it is not enough simply to stand on the savanna, enjoying the view.

To the south of Kenya, prolonged socialist liberation movements stimulated more social engagement among writers but perhaps less so among ordinary whites. The South African pastoral cannon of the nineteenth and early twentieth centuries had virtually ignored blacks. As J. M. Coetzee, the country's leading literary critic, writes. that literature presumed that "the ultimate fate of whites was going to depend ... on an accommodation with the South African landscape."[32] The interracial cultural landscape only broke into writers' awareness in 1960—the same year as Kenya's independence—when police killed peaceful demonstrators at Sharpeville. After that massacre, writers such as Nadine Gordimer criticized rather than reproduced mainstream white values. Yet those values did not immediately change. Her 1972 novel *The Conservationist* concerns a man "in love with his farm." The farm seems to reciprocate. "If you walk about this place on your own," the owner explains to his liberal activist girlfriend, "you see things you'd never see otherwise. Birds and animals—everything accepts you. But if you have [black] people tramping all over the place—."[33] Notwithstanding or perhaps because of Sharpeville, the character withdraws from South African society. Rhodesian whites responded in a similar fashion to the guerrilla war of the 1970s and independence in 1980. "Despite the years of war and upheaval, and the participation or involvement of the writers in it," notes the editor of a 1982 collection of poetry, "their preoccupation is very much with the mundane, and with Nature and the seasons of the land they loved."[34] What do black people have to do, one is tempted to ask, to hold whites' attention?[35]

They can give up crops or belong to a "primitive" group that has never farmed at all. As an exception to the rule, Africans who hunt and gather or herd cattle *have* captured whites' literary imagination.[36] Hunting narratives almost invariably associate loyal African trackers—from minority nonagricultural tribes—with equally devoted Great White Hunters.[37] An old

Shangaan tracker stalks Rhodesian game with the hero of Keith Meadows's novel *Sand in the Wind*.[38] They have known each other since the white man's boyhood, when "From his father and the wild tribal scarred black man who was his constant shadow he learned about the bush" (173). These personal histories and enduring bonds—exclusively male, linked to death, and spatially separate from society—run deeper than the human. Indeed, white writers engage most fully and directly with Africans represented as animalian. No one has generated more lasting and more widely accepted imagery of this kind than the Kalahari explorer and author Laurens van der Post. The San, he writes in an account of his 1952 expedition, consider wild mammals "as companions in mystery, as fellow pilgrims traveling on the same perilous spoor between life-giving waters." Passing from the faunal to the inanimate, "Wherever he [the Bushman] went he contained, and was contained deeply within, the symmetry of the land."[39] More recently, scholarship has established a different understanding of Kalahari peoples as cattle herders and long-distance traders.[40] Still, the primitivized forager figure allows whites to appreciate Africans without closing the social distance. "Klara . . . was a bushman woman, a Stone Age woman," recalls van der Post of his nanny. "I looked into her eyes and it was always as if I looked at the first dawn of the first day."[41] Beyond multiracialism and indeed beyond explicit racism as well, Klara serves as the environmental Other.

A thousand kilometers from Klara and the Kalahari, only South Africa's coast seems to enable a more thorough engagement with Africans. Perhaps coincidentally, Rooke and Coetzee, the authors mentioned above as addressing black-on-white violence, grew up in Natal and the Cape, respectively. Certainly ethnic hybridity marks the littoral's past and representations of that coast. In the eighteenth century, shipwrecked whites married into Xhosa chiefly lineages. Former guerrilla Hazel Crampton narrates these events with a feeling of personal liberation. Her popular history *The Sunburnt Queen* recalls Crampton's first visit to the Wild Coast while on leave from Umkhonto weSizwe, the armed resistance to apartheid. She hears of "a black guy with blue eyes" and due to her "apartheid-era schooling where history was all black and white and two never mixed," finds herself captivated.[42] For other white writers, the coast signals not peaceful integration but instead white extinction. Blacks might drive whites into the sea. Again, it nearly happened before. In 1856, the Xhosa prophetess Nongqawuse predicted a mass drowning of all colonials. "Perhaps Nongquase you will have your revenge," writes a contemporary South Africa poet, "a displaced people; our demise / Is near, and we'll be gutted where we fall."[43] Such counterhegemonic scenarios seem to separate the coast from the plateau. In Rian Malan's memoir *My Traitor's Heart*, the author's trekker forebear "left the Cape a racially enlightened man. And then he ... disappeared into Africa, where he was transformed as all white men who went there were transformed."[44] Cape Town—or, as van der Post puts it, "the arrogant political intellectualism of the Cape" (*Lost*, 81)—is the exception that proves the rule: whites sympathize with blacks and see past the landscape only outside of authentic Africa.

Thus, the coast and the interior present divergent forms of racial consciousness. Malan's Capetonians believe in integration. So did Portugal, whose "luso-tropicalismo" created mestizo populations in Maputo and Luanda.[45] Coastal colonials could also practice "heterophobia,"[46] fearing, hating, and wreaking violence upon the Other. Portugal permitted forced labor in its empire until 1961. In the nineteenth century "Cape liberals" propounded scientific racism, while contemporary Cape frontiersmen put those ideas into practice through enslavement and murder.[47] Even slavery required thought and consideration; slavery could stir in the master hate, love, or another emotional response. As Toni Morrison writes with respect to the United States, "Americans choose to talk about themselves through and within a sometimes allegorical, sometimes metaphorical, but always choked representation of an Africanist presence."[48] In the twentieth-century imaginary of Anglophone savanna settlers, however, there is an Africanist absence. While occupying and farming the interior, many whites adopted a posture of neither loving nor hating blacks but instead simply not caring much about them. Of course, colonial administrators wreaked every form of violence upon blacks, and settlers offhandedly described them as lazy, improvident, and incompetent.[49] Yet in the creative discourse with which whites portrayed themselves and their place in Africa, blacks bulked small; the land, plants, and animals bulked large. Jean and John Comaroff refer to this structure of feeling in 1990s' South Africa as "post-racist racism."[50] Put slightly differently, images of nature provided an escape route from racism *and* from multiracialism. In the minds of many whites, aesthetics of wilderness took the place of both ethnic chauvinism and pluralism. Landscape, then, served as an attractive though largely unrecognized third way. At the threshold of consciousness—just detectable by writers—whites ran from blacks and hid in the bush.

Embracing African Land

If whites found themselves by losing themselves in the bush, belonging did not come easily there either. Initially African landscapes unnerved whites

almost as much as did African people. The "thousand miles of Africa re-
ceding" from the Elkington's garden held few features reminiscent of England. Glaciers and year-round mild rain had sculpted the British Isles as an
intricate pattern of pond, shoreline, loch, and coast. Water interlacing with
soil distinguishes this topography and the art and literature of it. Think
of William Wordsworth's highly specific praise of the Lake District, where
he prefers a shoreline "gracefully or boldly indented."[51] East and southern
Africa, by contrast, hold little standing water (apart from the lakes of the
Great Rift Valley). Rain falls intermittently in torrents, racing almost unchecked to the Indian Ocean. The white immigrant would have to adapt.
The literate immigrant would have to do more: set aside and replace aesthetic sensibilities, Wordsworth, and much of the pastoral canon.[52] White
writers confronted a challenge later recognized by J. M. Coetzee, who asks,
"How are we to read the African landscape? Is it readable at all? Is it readable only through African eyes, writable only in an African language?" (62).
Yes, it was, as these writers demonstrated, but only with difficulty and with
the right coping mechanisms.

These mechanisms were threefold: an emphasis on glaciers, an apology
for color and form, and an embrace of Africa's vastness. The first technique
drew attention to those features most reminiscent of Europe's geology. Describing Kenya's Great Rift Valley, for instance, the early settler Frederick
de Janzé invokes "the glacial ethereal beauty of the valley."[53] With less need
for hyperbole, glaciophiles found the real object of their desire on Mount
Kilimanjaro. At 5,895 meters above sea level, the peak sustains year-round
ice. In 1927 Hemingway's "Snows of Kilimanjaro" began a tradition of gra-
tuitous reference to the peak. Even later works only tangentially related to
"Kili" mention it prominently or display it on the cover. The physician Michael Latham, for instance, rather incongruously titles his family memoir
Kilimanjaro Tales: The Saga of a Medical Family in Africa.[54] If Kilimanjaro
deflects writers from their proper subjects, it also forms the subject of rich
photo-literary archive. Those representations often emphasize the mountain's cosmopolitan quality. A climb to the summit, as one coffee-table
book suggests, takes a person "from the tropics to the Arctic." En route the
climber encounters temperate "high moors" and "heathland."[55] Symbolically, Africa metamorphoses at its highest point into European terrain. Yet
if the Scot weary of Africa can find refuge on these slopes, so can a Euro-
African completely at home on the continent. David Western, the maverick
conservationist who introduced community-based wildlife management to
Kenya, still exaggerates when considering the mountain. In his autobiography *In the Dust of Kilimanjaro,* he recalls that his "earliest memory . . . is of

Mount Kilimanjaro seen from the train window across an endless stretch of thorn country." Much later, a glimpse of the peak "perfectly matched my image of the archetypal Africa."[56] Kilimanjaro, then, represents Europe to some and Africa to others and, perhaps most conveniently for whites, the marriage of the two.

Whites located far from Kili and other snowy peaks fell back on a second approach, an attitude of apology, faint praise, and criticism vis-à-vis the savanna. In 1914 Steward Gore-Brown trekked to Northern Rhodesia to construct an elaborate manor on the remote, tiny Lake Shiwa Ngandu. The place spoke to him, or, as he inscribed on the mantle piece, "This corner of the earth, smiles on me more than any other." Yet a letter home described Shiwa Ngandu as a mere replica: "It looks like a little bit of Italy transported here to the middle of Africa. It's a Mediterranean colour, not an African blue in the least."[57] Color figures in another backhanded compliment. A review of Robert Paul, who painted Rhodesian landscapes from the 1940s onward, celebrates the way in which he could "set up an easel in front of a featureless expanse of nondescript grass and scrubby bushes, and with a few brushstrokes make something that one can look at endlessly. He found form, cohesion, variety, vitality in that seeming nothingness."[58]

Even "nothingness" could hold positive potential. The National Federation of Women's Institutes titled a 1967 publication for immigrants *Great Spaces Washed with Sun.*[59] Recalling the trope of empty land, the advertisement suggested a new aesthetic sensibility for savanna. Yet the authors could not maintain that sensibility through the narrative. As if losing their nerve, they revert to English standards and almost apologize for their landscape: "There are no great natural lakes in Rhodesia, and at times it is difficult for the people of *more favored countries* to realise what man-made lakes and dams mean to Rhodesians."[60] The same ambivalence surfaces in Lisa Fugard's recent novel *Skinner's Drift.* One of the main characters, Lorraine, "let the beauty of the farm at dusk break her heart." Later, however, drunk at party, she wishes for a fountain on her Limpopo Valley estate: "Yes, at night I'd lie in bed and listen to the play of the water instead of the nightjars. I know I am doomed to failure, but I keep on trying to civilize this part of the world."[61] Whites could be demanding parents to the savanna, constantly finding fault in it rather than simply loving it for what it was.

Only a third and more innovative sensibility found value in the grassland as it was. Skillful writers promoted the "endless stretch of thorn country" from the background into the foreground. The setting virtually became a character itself. Among Zimbabweans, John Gordon Davis, an adventure novelist, arguably pioneered this transformation. Writing in 1967, he describes the savanna floridly as "vast green empty spacious great foreverness, pregnant and primitive and exciting and virgin and even dangerous under the great vast blue sky."[62] Davis's tone evokes a sense of extraordinary and out-of-scale rather than conventional beauty. Kenya's settlers seem to have understood this aesthetic of space earlier and almost intuitively. Describing their dissolute, adulterous lifestyle of the 1930s, the journalist James Fox conjectures that "such grandiose surroundings were irresistible to the English settlers and often went to their heads." "*Folie de grandeur*" gripped many an otherwise faithful spouse.[63] Beyond such sexual restlessness, the surrounds stimulated a range of emotions. In *Flame Trees*, Huxley reports a conversation at sunset outside Thika: "'Yes, it is the sort of sky that angry Valkyries might ride across,' Ian agreed . . . 'it has a barbaric splendour in it, and an element of terror'" (127). If dusk provokes fear, it can equally well suggest liberation. As Gallman prepares to leave Italy for Kenya, she daydreams of "unbounded freedom, of wide open horizons and red sunsets. . . . I could smell the dry grasses of unknown savannahs" (*I Dreamed of Africa*, 17). Topographical openness implied—as any seemingly empty space would—both autonomy and vulnerability. White writers played with this opposition, but those most interested in settlement emphasized the emancipatory effect of savanna.

Aerial views provided still further reassurance, channeling settlers' aesthetic logic toward property claims. Early white Kenyans availed themselves of unparalleled access to light aircraft. Karen Blixen flew more than most thanks to her love affair with amateur pilot Denys Finch Hatton. She recalls in *Out of Africa* that "in the air, you are taken into the full freedom of the three dimensions; after long ages of exile and dream the homesick heart throws itself into the arms of space" (230). On a wing, the settler finds his or her place. Beryl Markham, a contemporary of Blixen and business partner of her husband, describes a similar airborne homecoming, notwithstanding the exotic. In *West with the Night* she writes that "Africa is mystic; it is wild; it is a sweltering inferno. . . . To a lot of people, as to myself, it is just 'home'" (8). Crisscrossing British Africa as a professional pilot, Markham developed a sensibility toward the savanna that was almost proprietary. "I . . . take off into the night," she reminisces, "Ahead of me lies a land that is unknown to the rest of the world and *only vaguely known to the African*—a strange mixture of grasslands, scrub, desert sand like long waves of the southern ocean" (15, my emphasis). Beyond such symbolic claims—and in the longer term—white landownership benefited materially from the aerial view. In Rhodesia and later Zimbabwe, commercial farmers planned crops from air photos, and, of course, government departments mapped almost without pause.[64] Constantly published and reproduced, this bird's-eye view combined aesthetics of the "tourist gaze" with an "imperial visibility."[65] Flight, in short, gave white writers and white farmers the means to develop and implement their newfound appreciation for the savanna.

As that sensibility quickened, whites explored its limits. Whites found that outside the savanna, perspectives could extend too far and angles open too wide. The Kalahari and adjoining deserts, for instance, exceeded most whites' capacity to adapt and embrace. John Gordon Davis describes ex-German Namibia—the setting for his combined treasure- and Nazi-hunting tale—as "the land God made in anger."[66] Namibian German geologist Henno Martin blames a related deity. "Inconceivable under a more temperate sky and in milder latitudes," he writes of the canyons where he hid from 1940 to 1942 to avoid interment, "the Devil had created them in an idle hour."[67] "As long as the word nature conjured up the green woods and the flower-strewn meadows of our childhood," Martin recalls, he and his companion can see neither beauty nor balance in the desert (117). Perhaps they should have exercised greater poetic license, as does Zimbabwe-born Michael Main in a photo book on the Kalahari. Main begins with geological data: the Magkagikgadi Pans are a paleolake that has dried out completely over the past fifty thousand years. Today only the largest floods deposit a thin sheet of water.[68] This empirical recounting soon gives way to the romance of anachronistic space: on (so-called) Kubu Island, there is "a pebble beach and it speaks of a time long, long ago when the pan was a sea. . . . There were times when the whole island was deep under water and others when it barely showed, and there were times, like today, when it lay bare and naked to the sun, an island in a forgotten sea" (26). The evocation plays with tense, conjuring into actuality a long-gone waterscape. In a pinch, then, literary whites could imagine away the driest and most unobstructed topographies without embracing them for what they were.

Landscapes at the opposite extreme—with fully obstructed lines of sight—created an insoluble problem. Having assimilated to the savanna, whites seemed to dread dense, closed woodland, and they sometimes responded to it in lethal ways. The Mau Mau Revolt, after all, took place in the Aberdare forest. "In a land where limitless space was supposed to be as much a part of your life as the furniture," Ruark writes, Valerie "had lost her husband to the black forests where he had gone to hunt men." Dermott, the husband, began his counterinsurgency patrols as a healthy soldier and finished them as a disturbed alcoholic (22–23). To the south, only the jesse forest of the Zambezi Valley causes anywhere near such distress. "The tracking was difficult," writes Zimbabwean Keith Meadows in a passage on hunting, in "that dense, tangled impenetrable barrier of cobretum that favours only the elephants and rhinos."[69] The forest transmits danger, terror, and evil, constituting what Taussig, describing torture in colonial Peru, calls a "space of death."[70] Regarding Africa, only Joseph Conrad—who wrote about central Africa but certainly influenced the settler colonies—conveys the full force of this zone.[71] His descriptions of the Congo rain forest draw attention to two features: visual opacity and the lurking presence of natives. "The great wall of vegetation . . . was like a rioting invasion of soundless life, a rolling wave of plants, piled up, crested, ready to topple over the creek to sweep every little man of us out his little existence."[72] More threatening still, the forest teems with Africans: "suddenly . . . I made out deep in the tangled gloom, naked breasts, arms, legs, glaring eyes—the bush was swarming with human limbs in movement" (125). Like Aberdare, this forest drives its European interloper mad. Dense inhabited vegetation, in other words, could unhinge colonials and colonialism from the inside out. Fortunately for all concerned, the more numerous whites on the savanna felt safe to ignore, rather than slaughter, Africans.

Thus, whites and the writers among them converted an uninviting topography into hospitable terrain. They could have borrowed from African meanings—such as the Zimbabwean oral history of "Guruuswa" (which means "long grass")[73]—and formed a syncretic sense of belonging. To do so would have required a fuller engagement with African Others. Rather than handle such strange materials, whites reworked European aesthetic sensibilities. They could not altogether overcome ambivalence, manifest in an obsession with montane anomalies and occasional faint praise for grassland. Nevertheless, Blixen and others effectively recalibrated the ideal of moderation to accord with African features. They replaced the scale of wet and dry, lake and upland, with a metric of open and closed. Glaciers, past or present, were no longer necessary. If the shoreline between land and water carried meaning before, the line of sight now served that function. Moderation and mildness inhered in vistas and horizons of the middle range, between the infinite and the overly finite. The savanna offered perspective and viewpoints, a less formal version of Renaissance planners' "geometry of landscape."[74] Like a panopticon, the savanna also revealed the presence of blacks or, more often due to colonial evictions, their comforting absence. By contrast, the tight angles and narrow interstices of forest could conceal Africans. At least the hunter, trader, farmer, or soldier could not forget them there. On open grass, these fears receded.

In this circumscribed fashion, whites' imaginative project took root in the last quarter of the twentieth century. Even as blacks were throwing off colonialism, whites *found* themselves on African savanna.

They did not do so securely or for long. Since 2000, violence against farmers in both Zimbabwe and South Africa has caused Euro-Africans to reassess their situation. Politics, at last, overwhelmed the imagination. A white Zimbabwean displaced in New York asks Douglas Rogers, "Do you ever think our ancestors got on the wrong boat?" Rogers, also a white Zimbabwean New Yorker, considers the question in relation to his Euro-American in-laws: "a simple quirk of geography meant they lived wildly different lives: *their* ancestors had taken different boats" (156, emphasis in original). If simple, the quirk is not a happy one. Englishmen disemarking in the New World in the eighteenth and nineteenth centuries wiped out and replaced native peoples. So complete was this erasure that conservationist Wes Jackson can title a book of essays *Becoming Native to This Place*.[75] In the United States, Anglophone whites mostly win contests of authenticity, asserting their history and language as quintessential. In Africa, by contrast, the current politics of autochthony and citizenship have dismissed whites easily and immediately.[76] Now more than ever, African whites appear to themselves and others as a "lost tribe." Uncomfortable and unsafe as it is, there is a certain honesty and humility in this position. When I first did ethnography fieldwork in Chimanimani District, Zimbabwe, whites frequently queried me, anxiously, on the meaning of the place-name. Born and "found" in Massachusetts, I only experienced anxiety when trying to spell the aboriginal word. It meant my home. Passage on the "right boat" often confers a self-assuredness bordering on ethnic entitlement. The imaginative work of white Africans—who never stopped trying to find themselves—helps rather better to build a plural society.

Notes

The epigraphs at the start of the chapter are from Isak Dinesen, *Out of Africa* (New York: Random House, 1937), 3–4, and Beryl Markham, *West with the Night* (Boston: Houghton Mifflin, 1942), 60.

1. Riccardo Orizio, *Lost White Tribes: The End of Privilege and the Last Colonials in Sri Lanka, Jamaica, Brazil, Haiti, Namibia, and Guadeloupe* (New York: Free Press, 2000), and Douglas Rogers, *The Last Resort: A Zimbabwe Memoir* (New York: Random House, 2009), 122.

2. Hughes, David McDermott, *Whiteness in Zimbabwe: Race, Landscape, and the Problem of Belonging* (New York: Palgrave Macmillan, 2010).

3. Edward W. Said, *Culture and Imperialism* (New York: Knopf, 1993), 14. Regarding "structures of feeling," see Raymond Williams, *Marxism and Literature* (Oxford: Oxford University Press, 1977), 132–35.

4. Dane Kennedy, *Islands of White: Settler Society and Culture in Kenya and Southern Rhodesia, 1890–1939* (Durham, NC: Duke University Press, 1987), 2.

5. Vincent Crapanzano, *Waiting: The Whites of South Africa* (New York: Vintage, 1986), xiv–xv; Peter Godwin and Ian Hancock, *"Rhodesians Never Die": The Impact of War and Political Change on White Rhodesia, c. 1970–1980* (Oxford: Oxford University Press,1993), 287; and Kennedy, *Islands*, 1.

6. Jill Lepore, *The Name of War: King Philip's War and the Origins of American Identity* (New York: Vintage, 1998).

7. Paul Rabinow, *French Modern: Norms and Forms of the Social Environment* (Cambridge, MA: MIT Press, 1989), 294, and Ann Laura Stoler, "Affective States," in *A Companion to the Anthropology of Politics*, edited by David Nugent and Joan Vincent, 4–20 (Oxford, UK: Blackwell, 2004).

8. Diana Jeater, "Speaking Like a Native: Vernacular Languages and the State in Southern Rhodesia, 1890–1935," *Journal of African History* 42, no. 3 (2001): 449–68.

9. Up to that point, Rhodesian writers frequently set novels in the context of the 1896–97 native uprisings or imaginary revolts. Anthony Chennells, "White Rhodesian Nationalism—the Mistaken Years," in *Turmoil and Tenacity: Zimbabwe, 1980–1990*, edited by Canaan S. Banana. (Harare: College Press, 1989), 124–25, 134.

10. Ngugi wa Thiong'o, *Detained: A Writer's Prison Diary* (Oxford, UK: Heinemann, 1981), 29.

11. Donald S. Moore, Jake Kosek, and Anand Pandian, eds., *Race, Nature, and the Politics of Difference* (Durham, NC: Duke University Press, 2003), 1.

12. George M. Fredrickson, *White Supremacy: A Comparative Study in American and South African History* (Oxford: Oxford University Press, 1981), 4, my emphasis.

13. Richard White, *The Middle Ground: Indians, Empires, and Republics in the Great Lakes Region, 1650–1815* (Cambridge: Cambridge University Press, 1991).

14. Allen Isaacman and Barbara Isaacman, "The Prazeiros as Transfrontiersmen: A Study in Social and Cultural Change," *International Journal of African Historical Studies* 8, no. 1 (1975): 1–39.

15. Doris Lessing, "The Old Chief Mshlanga," in Doris Lessing, *African Stories* (New York: Ballantine, 1951), 56, 58.

16. Terence Ranger, *Revolt in Southern Rhodesia, 1896–97: A Study in African Resistance* (Evanston, IL: Northwestern University Press, 1967), and Kennedy, *Islands*, 129.

17. J. M. Burns, *Flickering Shadows: Cinema and Identity in Colonial Zimbabwe* (Athens: Ohio University Press, 2002), and Ann Laura Stoler, *Carnal Knowledge and Imperial Power: Race and the Intimate in Colonial Rule* (Berkeley: University of California Press, 2002), 59–60.

18. Daphne Rooke, *Mittee* (New York: Houghton Mifflin, 1952).

19. Doris Lessing, *The Grass Is Singing* (New York: Crowell, 1950).

20. Peter Armstrong in his war novel *Operation Zambezi: The Raid into Zambia* (Salisbury, Rhodesia: Welston, 1979) recalls guerrillas' downing of a civilian passenger plane in 1978, regarding which "there was evidence to suggest that female survivors had been raped" upon crash-landing (5). Although guerrillas did shoot surviving passengers, Armstrong appears to have invented the sexual assault; see Godwin and Hancock, *"Rhodesians Never Die,"* 228.

21. Nadine Gordimer, "Living in the Interregnum," *New York Review of Books* 29, nos. 21/22 (1983): 21–29; cf. Crapanzano, *Waiting*, 39.

22. Paul Carter, *The Road to Botany Bay: An Exploration of Landscape and History* (New York: Knopf, 1987).

23. William Cronon, *Changes in the Land: Indians, Colonists, and the Ecology of New England* (New York: Hill and Wang, 1983), 57.

24. Elspeth Huxley, *The Flame Trees of Thika: Memories of an African Childhood* (Harmondsworth, UK: Penguin, 1959), 222.

25. Earlier, in fact, Huxley had written a novel, *Red Strangers*, exclusively about Kikuyu, but she understood that work as exceptional. Later the author described *Red Strangers* as "a foolhardy idea, since I doubt whether any member of one race can get under the skin of people of another race and culture"; see Elspeth Huxley, *Out in the Midday Sun* (Harmondsworth, UK: Penguin, 1985), 182.

26. Anne McClintock, *Imperial Leather: Race, Gender and Sexuality in the Colonial Conquest* (London: Routledge, 1995), 41, 30.

27. Maughan-Brown includes Ruark in his treatise on Kenyan literature due to his immense influence in and regarding the colony, his "oracular status in some circles in South Africa"; see David Maughan-Brown, *Land, Freedom, and Fiction: History and Ideology in Kenya* (London: Zed Books, 1985), 15.

28. Robert Ruark, *Uhuru* (London: Corgi, 1962), 310, emphasis in original.

29. Kuki Gallman, *I Dreamed of Africa* (London: Penguin, 1991), 229.

30. Francesca Marciano, *Rules of the Wild* (London: Vintage, 1999).

31. Gus Le Breton, personal communication, Harare, May 15, 2006.

32. J. M. Coetzee, *White Writing: On the Culture of Letters in South Africa* (New Haven, CT: Yale University Press, 1988), 8.

33. Nadine Gordimer, *The Conservationist* (London: Penguin, 1972), 176. Caminero-Santangelo (this volume) discusses the novel in much greater depth.

176 | *David McDermott Hughes*

34. Louis W. Bolze, "Publisher's Introduction," in Rowland Molony, David Wright, John Eppel, and Noel Brettell, *Four Voices: Poetry from Zimbabwe* (Bulawayo, Zimbabwe: Books of Zimbabwe, 1982), x.

35. In fact, the war did spawn a series of macho tales of cross-racial violence, most of which were quickly forgotten. See Armstrong, *Operation Zambezi*; Robert Early, *A Time of Madness* (Salisbury, Rhodesia: Graham Publishing, 1977); and Antony Trew, *Towards the Tamarind Trees* (London: Collins, 1970).

36. The smaller number of Africans who kill people for a living have earned the same distinction. In modern armies, black warriors and white warriors have broken all the rules of segregation and identity. See Luise White, "Precarious Conditions: A Note on Counter-Insurgency in Africa after 1945," *Gender and History* 16, no. 3 (2004): 603–25.

37. In Kenya, pastoralists infected Europeans with "'Masai-itis,' an emotional obsession with the Masai," which destroyed their desire to rule. See Kathryn Tidrick, *Empire and the English Character* (London: I. B. Tauris, 1990), 172–73; cf. Dorothy Hodgson, *Once Intrepid Warriors: Gender, Ethnicity, and the Cultural Politics of Maasai Development* (Bloomington: Indiana University Press, 2002), 2. The Maasai *moran,* or male warrior, can still stimulate another desire, as recorded by the German ex-wife of one such individual; see Corinne Hofmann, *The White Masai,* translated by Peter Millar (London: Arcadia Books, 2005 [1998]).

38. Keith Meadows, *Sand in the Wind* (Bulawayo, Zimbabwe: Thorntree, 1996).

39. Laurens van der Post, *The Lost World of the Kalahari* (New York: Morrow, 1958), 21; cf. Edwin N. Wilmsen, "Primitive Politics in Sanctified Landscapes: The Ethnographic Fictions of Laurens van der Post," *Journal of Southern African Studies* 21, no. 2 (1995): 201–23.

40. Edwin N. Wilmsen, *Land Filled with Flies: A Political Economy of the Kalahari* (Chicago: University of Chicago Press, 1989), 98–101.

41. Laurens van der Post, *A Walk with a White Bushman* (New York: Morrow, 1986), 3.

42. Hazel Crampton, *The Sunburnt Queen: A True Story* (Johannesburg: Jacana Media, 2004), 15.

43. Christopher Hope, "The Flight of the White South Africans," In *A Land Apart: A Contemporary South African Reader,* edited by André Brink and J. M. Coetzee, 116–17 (New York: Penguin, 1986).

44. Rian Malan, *My Traitor's Heart* (London: Vintage, 1991), 21.

45. See Gilberto Freyre, "Man Situated in the Tropics: Metarace and Brown Skins," in *The Gilberto Freyre Reader,* translated by Barbara Shelby, 83–85 (New York: Knopf, 1974), and Brian Owensby, "Toward a History of Brazil's 'Cordial

Racism': Race beyond Liberalism," *Comparative Studies in Society and History* 47, no. 2 (2005): 318–47, for celebration and analysis of *lusotropicalismo,* respectively.

46. Albert Memmi, *Racism,* translated by Steve Martino (Minneapolis: University of Minnesota Press, 2000), 117–21.

47. Saul Dubow, "Earth History, Natural History, Prehistory at the Cape, 1860–1875," *Comparative Studies in Society and History* 46, no. 1 (2004): 133. Zine Magubane "Simians, Savages, Skulls, and Sex: Science and Colonial Milenarism in Nineteeth-Century South Africa," in *Race, Nature, and the Politics of Difference,* 99–121.

48. Toni Morrison, *Playing in the Dark: Whiteness and the Literary Imagination* (Cambridge: Harvard University Press, 1992), 17.

49. Blair Rutherford, *Working on the Margins: Black Workers, White Farmers in Postcolonial Zimbabwe* (London: Zed Books, 2001), 88–89.

50. Jean Comaroff and John Comaroff, "Naturing the Nation: Aliens, Apocalypse and the Postcolonial State," *Journal of Southern African Studies* 27, no. 3 (2001): 627–51.

51. William Wordsworth, *A Description of the Lakes* (Oxford, UK: Woodstock Books, 1991 [1822]), 23.

52. Raymond Williams, *The Country and the City* (Oxford: Oxford University Press, 1973).

53. Le Comte de Janzé, *Vertical Land* (London: Duckworth, 1928), 65.

54. Gwynneth Latham and Michael Latham, *Kilimanjaro Tales: The Saga of a Medical Family in Africa* (London: Radcliffe Press, 1995).

55. David Pluth, Mohamed Amin, and Graham Mercer, *Kilimanjaro: The Great White Mountain of Africa* (Nairobi: Camerapix, 2001), 94, 115.

56. David Western, *In the Dust of Kilimanjaro* (Washington, DC: Island Press, 1997), 7.

57. Christina Lamb, *The Africa House* (London: Viking, 1999), frontispiece, 135. Gore-Brown actually wrote the inscription in Latin as "*Ille terrarum mihi super omnes angulus ridet.*"

58. François Roux, "Clarity and Discipline: An Appreciation of Robert Paul," in *Robert Paul,* edited by Barbara Murray (Harare: Colette Wiles, 1996), 60.

59. Kipling used this phrase in his elegy to Cecil Rhodes, "The Burial," read at Rhodes's funeral in the Matopos in 1902; see Terence Ranger, "'Great Spaces Washed with Sun': The Matopos and Uluru Compared," In *Text, Theory, Space: Land, Literature, and History in South Africa and Australia,* edited by Kate Darian-Smith, Liz Gunner, and Sarah Nuttall (London: Routledge, 1996), 169n1.

60. National Federation of Women's Institutes, *Great Spaces Washed with Sun* (Salisbury, Rhodesia: M. O. Collins, 1967), 188, my emphasis. See also David

McDermott Hughes, "Whites and Water: How Euro-Africans Made Nature at Kariba Dam," *Journal of Southern African Studies* 32, no. 4 (2006): 823–38.

61. Lisa Fugard, *Skinner's Drift* (New York: Scribner, 2006), 147, 224.

62. John Gordon Davis, *Hold My Hand I'm Dying* (London: Michael Joseph, 1967), 301.

63. James Fox, *White Mischief: The Murder of Lord Erroll* (New York: Vintage, 1982).

64. David McDermott Hughes, *From Enslavement to Environmentalism: Politics on a Southern African Border* (Seattle: University of Washington Press, 2006), 70.

65. John Urry, *The Tourist Gaze: Leisure and Travel in Contemporary Societies* (London: Sage, 1990), and D. Graham Burnett, *Masters of All They Surveyed: Exploration, Geography, and a British El Dorado* (Chicago: University of Chicago Press, 2000), 126–29.

66. John Gordon Davis, *The Land God Made in Anger* (Glasgow: Collins, 1990), 6.

67. Henno Martin, *The Sheltering Desert* (Jeppestown, South Africa: Donker, 1983), 20–21.

68. Michael Main, *Kalahari: Life's Variety in Dune and Delta* (Johannesburg: Southern Book Publishers, 1987), 18.

69. Keith Meadows, *Rupert Fothergill: Bridging a Conservation Era* (Bulawayo, Zimbabwe: Thorntree, 1981), 71.

70. Michael Taussig, *Shamanism, Colonialism, and the Wild Man: A Study in Terror and Healing* (Chicago: University of Chicago Press, 1987), 5, 78.

71. See David Ward, *Chronicles of Darkness* (London: Routledge, 1989), for a description of Conradian themes in white writing of the savanna.

72. Joseph Conrad, "Heart of Darkness," in *Youth: A Narrative and Two Other Stories* (Edinburgh, UK: William Blackwood, 1902), 98.

73. D. N. Beach, *The Shona and Their Neighbors, 900–1850* (Gweru, Zimbabwe: Mambo, 1980), 62–63.

74. Denis Cosgrove, "The Geometry of Landscape: Practical and Speculative Arts in Sixteenth-century Venetian Land Territories," in *The Iconography of Landscape*, edited by Denis Cosgrove and Stephen Daniels, 254–76 (Cambridge: Cambridge University Press, 1988).

75. Wes Jackson, *Becoming Native to This Place* (New York: Counterpoint, 1994).

76. Peter Geschiere, *The Perils of Belonging: Autochthony, Citizenship, and Exclusion in Africa and Europe* (Chicago: University of Chicago Press, 2009).

Bibliography

Armstrong, Peter. *Operation Zambezi: The Raid into Zambia.* Salisbury, Rhodesia: Welston, 1979.

Beach, D. N. *The Shona and Their Neighbors, 900–1850.* Gweru, Zimbabwe: Mambo, 1980.

Bolze, Louis W. "Publisher's Introduction." In Rowland Molony, David Wright, John Eppel, and Noel Brettell, *Four Voices: Poetry from Zimbabwe,* ix–x. Bulawayo, Zimbabwe: Books of Zimbabwe, 1982.

Burnett, D. Graham. *Masters of All They Surveyed: Exploration, Geography, and a British El Dorado.* Chicago: University of Chicago Press, 2000.

Burns, J. M. *Flickering Shadows: Cinema and Identity in Colonial Zimbabwe.* Athens: Ohio University Press, 2002.

Carter, Paul. *The Road to Botany Bay: An Exploration of Landscape and History.* New York: Knopf, 1987.

Chennells, Anthony. "White Rhodesian Nationalism—the Mistaken Years." In *Turmoil and Tenacity: Zimbabwe, 1980–1990,* edited by Canaan S. Banana, 123–39. Harare: College Press, 1989.

Coetzee, J. M. *White Writing: On the Culture of Letters in South Africa.* New Haven, CT: Yale University Press, 1988.

Comaroff, Jean, and John Comaroff. "Naturing the Nation: Aliens, Apocalypse and the Postcolonial State." *Journal of Southern African Studies* 27, no. 3 (2001): 627–51.

Conrad, Joseph. "Heart of Darkness." In *Youth: A Narrative and Two Other Stories,* 49–182. Edinburgh, UK: William Blackwood, 1902.

Cosgrove, Denis. "The Geometry of Landscape: Practical and Speculative Arts in Sixteenth-century Venetian Land Territories." In *The Iconography of Landscape,* edited by Denis Cosgrove and Stephen Daniels, 254–76. Cambridge: Cambridge University Press, 1988.

Crampton, Hazel. *The Sunburnt Queen: A True Story.* Johannesburg: Jacana Media, 2004.

Crapanzano, Vincent. *Waiting: The Whites of South Africa.* New York: Vintage, 1986.

Cronon, William. *Changes in the Land: Indians, Colonists, and the Ecology of New England.* New York: Hill and Wang, 1983.

Crosby, Alfred W. *Ecological Imperialism: The Biological Expansion of Europe, 900–1900.* Cambridge: Cambridge University Press, 1986.

Davis, John Gordon. *Hold My Hand I'm Dying.* London: Michael Joseph, 1967.

———. *The Land God Made in Anger.* Glasgow: Collins, 1990.

de Janzé, Le Comte de. *Vertical Land.* London: Duckworth, 1928.

Dinesen, Isak. *Out of Africa.* New York: Random House, 1937.

Dubow, Saul. "Earth History, Natural History, Prehistory at the Cape, 1860–1875." *Comparative Studies in Society and History* 46, no. 1 (2004): 107–33.

Early, Robert. *A Time of Madness.* Salisbury, Rhodesia: Graham Publishing, 1977.

Fugard, Lisa. *Skinner's Drift.* New York: Scribner, 2006.

Fox, James. *White Mischief: The Murder of Lord Erroll.* New York: Vintage, 1982.

Fredrickson, George M. *White Supremacy: A Comparative Study in American and South African History.* Oxford: Oxford University Press, 1981.

Freyre, Gilberto. "Man Situated in the Tropics: Metarace and Brown Skins." In *The Gilberto Freyre Reader,* translated by Barbara Shelby, 83–85. New York: Knopf, 1974.

Gallman, Kuki. *I Dreamed of Africa.* London: Penguin, 1991.

Geschiere, Peter. *The Perils of Belonging: Autochthony, Citizenship, and Exclusion in Africa and Europe.* Chicago: University of Chicago Press, 2009.

Gilroy, Paul. *Between Camps: Nations, Cultures, and the Allure of Race.* London: Routledge, 2000.

———. *The Black Atlantic: Modernity and Double Consciousness.* Cambridge: Harvard University Press, 1993.

Godwin, Peter, and Ian Hancock. *"Rhodesians Never Die": The Impact of War and Political Change on White Rhodesia, c. 1970–1980.* Oxford: Oxford University Press, 1993.

Gordimer, Nadine. *The Conservationist.* London: Penguin, 1972.

———. *July's People.* New York: Viking, 1981.

———. "Living in the Interregnum." *New York Review of Books* 29, nos. 21/22 (1983): 21–29.

Hemingway, Ernest. *Snows of Kilimanjaro and Other Stories.* New York: Scribner, 1927.

Hodgson, Dorothy. *Once Intrepid Warriors: Gender, Ethnicity, and the Cultural Politics of Maasai Development.* Bloomington: Indiana University Press, 2001.

Hofmann, Corinne. *The White Masai.* Translated by Peter Millar. London: Arcadia Books, 2005 [1998].

Hope, Christopher. "The Flight of the White South Africans." In *A Land Apart: A Contemporary South African Reader,* edited by André Brink and J.M. Coetzee, 116–17. New York: Penguin, 1986.

Hughes, David McDermott. *From Enslavement to Environmentalism: Politics on a Southern African Border.* Seattle: University of Washington Press, 2006.

———. *Whiteness in Zimbabwe: Race, Landscape, and the Problem of Belonging.* New York: Palgrave Macmillan, 2010.

———. "Whites and Water: How Euro-Africans Made Nature at Kariba Dam." *Journal of Southern African Studies* 32, no. 4 (2006): 823–38.

Huxley, Elspeth. *The Flame Trees of Thika: Memories of an African Childhood.* Harmondsworth, UK: Penguin, 1959.

———. *Out in the Midday Sun.* Harmondsworth, UK: Penguin, 1985.

———. *Red Strangers.* London: Chatto, 1939.

Isaacman, Allen, and Barbara Isaacman. "The Prazeiros as Transfrontiersmen: A Study in Social and Cultural Change." *International Journal of African Historical Studies* 8, no. 1 (1975): 1–39.

Jackson, Wes. *Becoming Native to This Place.* New York: Counterpoint, 1994.

Jeater, Diana. "Speaking Like a Native: Vernacular Languages and the State in Southern Rhodesia, 1890–1935." *Journal of African History* 42, no. 3 (2001): 449–68.

Kennedy, Dane. *Islands of White: Settler Society and Culture in Kenya and Southern Rhodesia, 1890–1939.* Durham, NC: Duke University Press, 1987.

Kipling, Rudyard. *Rudyard Kipling's Verse, Inclusive Edition, 1885–1926.* Garden City, NY: Doubleday, 1928.

Lamb, Christina. *The Africa House.* London: Viking, 1999.

Latham, Gwynneth, and Michael Latham. *Kilimanjaro Tales: The Saga of a Medical Family in Africa.* London: Radcliffe Press, 1995.

Lepore, Jill. *The Name of War: King Philip's War and the Origins of American Identity.* New York: Vintage, 1998.

Lessing, Doris. *The Grass Is Singing.* New York: Crowell, 1950.

———. "The Old Chief Mshlanga." In Doris Lessing, *African Stories,* 49–60. New York; Ballantine, 1951.

Main, Michael. *Kalahari: Life's Variety in Dune and Delta.* Johannesburg: Southern Book Publishers, 1987.

Malan, Rian. *My Traitor's Heart.* London: Vintage, 1991.

Marciano, Francesca. *Rules of the Wild.* London: Vintage, 1999.

Magubane, Zine. "Simians, Savages, Skulls, and Sex: Science and Colonial Milenarism in Nineteenth-Century South Africa." In *Race, Nature, and the Politics of Difference,* edited by Donald S. Moore, Jake Kosek, and Anand Pandian, 99–121 (Durham, NC: Duke University Press, 2003).

Markham, Beryl. *West with the Night.* Boston: Houghton Mifflin, 1942.

Martin, Henno. *The Sheltering Desert.* Jeppestown, South Africa: Donker, 1983.

Maughan-Brown, David. *Land, Freedom, and Fiction: History and Ideology in Kenya.* London: Zed Books, 1985.

McClintock, Anne. *Imperial Leather: Race, Gender and Sexuality in the Colonial Conquest.* London: Routledge, 1995

Meadows, Keith. *Rupert Fothergill: Bridging a Conservation Era.* Bulawayo, Zimbabwe: Thorntree, 1981.

———. *Sand in the Wind.* Bulawayo, Zimbabwe: Thorntree, 1996.

———. *Sometimes When It Rains: White Africans in Black Africa.* Bulawayo, Zimbabwe: Thorntree, 2000.

Memmi, Albert. *Racism.* Translated by Steve Martino. Minneapolis: University of Minnesota Press, 2000.

Moore, Donald S., Jake Kosek, and Anand Pandian, eds. *Race, Nature, and the Politics of Difference.* Durham, NC: Duke University Press, 2003.

Moore, Donald S., Anand Pandian, and Jake Kosek. "The Cultural Politics of Race and Nature: Terrains of Power and Practice." In *Race, Nature, and the Politics of Difference,* edited by Donald S. Moore, Jake Kosek, and Anand Pandian, 1–70. Durham, NC: Duke University Press, 2003.

Morrison, Toni. *Playing in the Dark: Whiteness and the Literary Imagination.* Cambridge: Harvard University Press, 1992.

National Federation of Women's Institutes. *Great Spaces Washed with Sun.* Salisbury, Rhodesia: M. O. Collins, 1997.

Ngugi wa Thiong'o. *Detained: A Writer's Prison Diary.* Oxford, UK: Heinemann, 1981.

Orizio, Riccardo. *Lost White Tribes: The End of Privilege and the Last Colonials in Sri Lanka, Jamaica, Brazil, Haiti, Namibia, and Guadeloupe.* New York: Free Press, 2000.

Owensby, Brian. "Toward a History of Brazil's 'Cordial Racism': Race beyond Liberalism." *Comparative Studies in Society and History* 47, no. 2 (2005): 318–47.

Pluth, David, Mohamed Amin, and Graham Mercer. *Kilimanjaro: The Great White Mountain of Africa.* Nairobi: Camerapix, 2001.

Rabinow, Paul. *French Modern: Norms and Forms of the Social Environment.* Cambridge, MA: MIT Press, 1989.

Ranger, Terence. "'Great Spaces Washed with Sun': The Matopos and Uluru Compared." In *Text, Theory, Space: Land, Literature, and History in South Africa and Australia,* edited by Kate Darian-Smith, Liz Gunner, and Sarah Nuttall, 157–71. London: Routledge, 1996.

————. *Revolt in Southern Rhodesia, 1896–97: A Study in African Resistance.* Evanston, IL: Northwestern University Press, 1967.

Rogers, Douglas. *The Last Resort: A Zimbabwe Memoir.* New York: Random House, 2009.

Rooke, Daphne. *Mittee.* New York: Houghton Mifflin, 1952.

Roux, François. "Clarity and Discipline: An Appreciation of Robert Paul." In *Robert Paul,* edited by Barbara Murray, 58–63. Harare: Colette Wiles, 1996.

Ruark, Robert. *Uhuru.* London: Corgi, 1962.

Rutherford, Blair. *Working on the Margins: Black Workers, White Farmers in Postcolonial Zimbabwe.* London: Zed Books, 2001. Working on the Margins: Black Workers, White Farmers in Postcolonial Zimbabwe. Harare: Weaver and London:

Said, Edward W. *Culture and Imperialism.* New York: Knopf, 1993.

————. *Orientalism.* New York: Pantheon, 1978.

Stoler, Ann Laura. "Affective States." In *A Companion to the Anthropology of Politics,* edited by David Nugent and Joan Vincent, 4–20. Oxford, UK: Blackwell, 2004.

————. *Carnal Knowledge and Imperial Power: Race and the Intimate in Colonial Rule.* Berkeley: University of California Press, 2002.

Taussig, Michael. *Shamanism, Colonialism, and the Wild Man: A Study in Terror and Healing.* Chicago: University of Chicago Press, 1987.

Tidrick, Kathryn. *Empire and the English Character.* London: I. B. Tauris, 1990.

Trew, Antony. *Towards the Tamarind Trees.* London: Collins, 1970.

Urry. John. *The Tourist Gaze: Leisure and Travel in Contemporary Societies.* London: Sage, 1990.

van der Post, Laurens. *The Lost World of the Kalahari.* New York: Morrow, 1958.

————. *A Walk with a White Bushman.* New York: Morrow, 1986.

Ward, David. *Chronicles of Darkness.* London: Routledge, 1989.

Western, David. *In the Dust of Kilimanjaro.* Washington, DC: Island Press, 1997.

White, Luise. "Precarious Conditions: A Note on Counter-Insurgency in Africa after 1945." *Gender and History* 16, no. 3 (2004): 603–25.

White, Richard. *The Middle Ground: Indians, Empires, and Republics in the Great Lakes Region, 1650–1815.* Cambridge: Cambridge University Press, 1991.

Williams, Raymond. *The Country and the City.* Oxford: Oxford University Press, 1973.

————. *Marxism and Literature.* Oxford: Oxford University Press, 1977.

Wilmsen, Edwin N. *Land Filled with Flies: A Political Economy of the Kalahari.* Chicago: University of Chicago Press, 1989.

————. "Primitive Politics in Sanctified Landscapes: The Ethnographic Fictions of Laurens van der Post." *Journal of Southern African Studies* 21, no. 2 (1995): 201–23.

Wordsworth, William. *A Description of the Lakes.* Oxford, UK: Woodstock Books, 1991 [1822].

Chapter 8

Waste and Postcolonial History

An Ecocritical Reading of J. M. Coetzee's
Age of Iron

Anthony Vital

IF ECOCRITICISM in the global North began with focus on the North's tradition of "Nature writing" and its associated environmental concerns, it has since reached into the North-South divide to articulate a more historically informed sense of nature and human need, with Rob Nixon's "Environmentalism and Postcolonialism" (2005) cited frequently as a marker of this shift.[1] Such ecocriticism, as it reaches into a global world, has to take into account this world's roots in European colonialism, with the accompanying social displacements, expropriations, and enslavements as well as the many forms of resistance, accommodation, and transformation that colonial activities have generated. Tracking the overall subordination of both people and nonhuman nature to an increasingly global economy, ecocriticism also has to acknowledge (without accepting) the current economy's power in providing material shelter for vast numbers of the planet's people and, through the work of such agencies as the United Nations and numerous nongovernmental organizations (NGOs), offering to many who are suffering crushing poverty a hope of refuge. Political and ethical questions follow from the recognition of such power—How are the comfortable, the prosperous, to live?—especially when the busy life of maintaining social position allows traces of this world's past violence (and its continuation, shaping the present) to appear only infrequently, briefly, as if through cracks in an otherwise ordinary and reliable surface.[2] A postcolonial ecocriticism might help answer such questions as it keeps its eye on this undercurrent of violence and develops interpretive modes able to connect the social behaviors that propelled the long phase of colonizing modernity (still under way and launched from multiple positions on the planet) with the accompanying damage to the earth's life systems and nonhuman species. Grasping too how modernity's energies have been institutionalized and reproduced, modified and refined, during this long stretch of time, such criticism would explore an appropriate politics and ethics.

Critical work of this sort, about and for Africa, would likely look different from criticism about and for North America, for example. Yet the novels of J. M. Coetzee, especially those written before the 1990s, suggest interpretive modes designed to narrow that distance. (His first published work of fiction, *Dusklands* [1974], stands as an obvious example, collecting two novellas, one set in the United States of America during the Vietnam War and the other set in eighteenth-century southern Africa). Developing postcolonial narrative forms, the novels frame their explorations with awareness of the complex operations of discourse and power linking metropole and colony in the expansion out of Europe that produced modernity's transformative force. Of the earlier novels, *Life & Times of Michael K* (1984 [1983]) has received ecocritical attention for being explicitly about the nature-human relation, with K, the gardener from Cape Town, seeking to escape the forms of administration and enclosure that mark modern societies globally. Yet while the novel puts into question Romantic attitudes toward nature and history, it still works within their structuring binaries and to that extent stimulates traditional associations with nature, recalling widespread ideas in the North about nature as "outside," as refuge from society. *Age of Iron* (1990), in contrast, develops a meditation on the nature-human relation that articulates an imagining of nature more richly appropriate to African worlds fallen under the domination of Europe-originated forms of modernity.[3]

To consider this later novel as being about nature appears counterintuitive: Mrs. Curren is very different from K, with her life in her large house in an urban landscape and with her professional background and, of course, in her articulateness as she puzzles in her lyrical voice over how to behave well in ugly times. *Age of Iron* seems to invite an ecocritical interpretation that reads the novel as symptomatic of an urban ideology unconcerned with a nature-human relation. Yet reading the novel's complex literary strategies in relation to a variety of political and environmental discourses, both of its own time and of ours, indicates its profound engagement with this relation. *Age of Iron*'s narrative, like *Michael K*'s, draws on literary discourses informing both a postmodern metafiction and realism, and it is, I argue, in the novel's realist dimension, which develops characters in identifiable environments, that *Age of Iron* proposes its complex insights into modern societies' exploitation of both people and land. The term that gives coherence to these insights is *waste,* one with a complex history (which I will touch on later) and connected in the text with a variety of recurring related terms, which together develop the narrative's evaluation of its material world. This writing of "waste" aligns the novel with discourses informing both the empirical study of waste in Africa and South Africa's current environmental justice writing, and thereby exposes blindnesses in a mainstream environmentalism, one that draws on images of nature that are silent about histories of colonizing.

If the realist dimension depicts a specific world, suggesting the relation of its structure to its past, then the novel's self-aware artifice reaches beyond this specificity to encourage generalized questions of meaning, thereby implicating all who read in its questioning. Critical response to the novel indicates the interpretive problems that such narrative poses as it suggests that the literal both matters *and* needs to be understood as subordinate to textuality, and valuable readings focusing on metafictional narration explore Curren's ethical quest.[4] Yet highlighting the narrative's representation of materiality discloses additional complexity in its exploration of ethics. Cape Town, emerging in detailed depiction as a world of waste, of what is cast out for being useless, unable to support a flourishing life, serves not only as impetus for Curren's ethical quest but also as comment on it. Curren's attempt, in opposition to this enveloping world, to find with her informed intelligence the right formulation, the right understanding, appears fragile, if not absurd, against the weight of a modern society perpetuating colonialism's social and material practices.[5] In this context, the narrative's final refusal of verisimilitude returns readers to kinds of question that hover over a postcolonial ecocriticism, questions of what it means to write, to try to live well, knowing that if we inherit the privileges afforded by modernity, these privileges, with the forms of knowing and sense of self that they offer, implicate us, however unwillingly, in modernity's distinctive damage to life.

It is of course through Curren's subjectivity that the narrative delivers its realist dimension, a subjectivity that, from the start, is marked by awareness of terminal illness. This awareness gives plausibility to Curren's noting, on her return home, an alley that she considers a "dead" place and also supplies plausibility to the narrative's emotional and imaginative intensities and their associated stylistic features. The narrator, driven to metaphor, reveals the materiality of people and place through an interplay of the literal and the figurative (Vercueil is both bad-smelling *and* someone she toys with thinking of as "angel" [4, 19, 14, 168]). Moreover, she complicates literal description by building sentences that accumulate appositive phrases and by repeating phrases in a variety of contexts across the narrative, thereby constructing overlapping patterns of association. The novel's opening paragraph, with its reference to "windblown leaves that pile up and rot" (3), both establishes the narrative's interest in "waste" and introduces this associative and metaphoric depictive style. While "windblown [and dead] leaves" in a narrative supposedly from a retired classics professor could bear intertextual reference to Virgil and Dante,[6] Curren saves the epithet "dead" for the "alley down the side of the garage"; she marks the leaves as rotting and thus points to literal organic decomposition. This initial naming of the alley links decomposition to death and uselessness, and the "windblown leaves" suggest litter as well as slow decomposition. These terms reappear as the narrative develops and can be linked to other terms, such as "rubbish," "refuse," "shit," and "derelict."

For ecocriticism, it is important to note that this reference to materiality does not draw on ecological discourse. While the "rotting" of leaves refers to a purely biological process, a focus on the organic alone is insufficient to signal ecological interest, for the ecological draws attention to the constitutive relation of organism and an environment comprising other species as well as nonliving material flows. *Age of Iron's* interest lies in decomposition generally (most obviously in Curren's own imminent organic failure) rather than in a vision of species interaction and environmental relation. But through the novel's poetic discursive texture and its plotting, what is biological develops a relation to social and cultural processes, the most obvious being the link that Curren makes between the cancer that prefigures her death to the forces bringing about the cultural and social disintegration around her. The novel thus develops a vision of the living-as-organic as subject to the rhythms of its own flourishing and decay but, more important, in historical time subject to social power flows sensed to be invasive and destructive. Reading this way for the narrative's representation of materiality, both literal and figurative, brings materiality back into view and gives interpretive context to the novel's epistemological and ethico-political concerns.

How the narrative uses materiality to situate these concerns can be seen in a brief moment during Curren's visit to Guguletu, one of many moments, moreover, that suggest that her predisposition to seeing "waste" increases rather than distorts her ability to observe. Here she witnesses state-sponsored vigilante action in the violent removal of shacks from the margins of this state-sponsored black township. When she has finished speaking in response to a request that she tell her opinion of what she sees, the narrative presents the following exchange:

> "This woman talks shit," said a man in the crowd. He looked around. "Shit," he said. No-one contradicted him. Already some were drifting away.
>
> "Yes," I said, speaking directly to him. "You are right, what you say is true."
>
> He gave me a look as if I were mad.
>
> "But what do you expect?" I went on. "To speak of this"—waving a hand over the bush, the smoke, the filth littering the path—"you would need the tongue of a god."
>
> "Shit," he said again, challenging me. (99)

The phrase "talking shit" serves here as an evaluative response complicating whatever justification the narrative may give for Curren's use of her high-flown phrase "tongue of a god." Let me clarify the obvious: *shit* is a term for solid excrement, material by-product of the body's need to ingest to support life, that the body discards. As with such colloquial, impolite usage and especially the four-letter kind, the word speaks to a troubled subjective response to what it denotes: anxiety, disgust, fear, etc., combining with fascination. "Talking" appears as "shit" when words cannot be ignored but appear repulsive, disgusting, the way feces appear when confronted unexpectedly in places where they do not belong. The phrase at this moment in the text serves two important functions: first, it rematerializes Curren's words, indicating a limit to her idealistic talk of gods, returning it to this particular material situation (in which she herself notes "filth"); second, it nudges readers to recognize the subjectivity of this nameless "man in the crowd" and his historically manufactured social and physical environments (the shacks, the smoke, the filth, the litter). He, and the people who share his world and this scene, would plausibly assess her words' value as "shit," both for being out of place, useless to them, and for being not simply useless, but repulsively so, for being evacuated, there in front of them, by someone from the social class bearing responsibility for their situation. That she finds "true" his first claim that her words are "shit" is sign of her increasing awareness of how social privilege has limited her subjectivity.[7]

Setting Curren's words in material context not only occurs in isolated incidents but also is built into the novel's narrative structure. Critics have noted that Curren, as character, is not static, yet no critics have noted that one key indicator of how she changes in response to experience is her shifting attitude toward home and the accompanying logically related attitude toward waste. Curren's awareness of pervasive waste remains constant, but her changing attitude becomes a marker of her moral development and feeds into the novel's ethical questioning. As homeowner, Curren at first feels intruded on by Vercueil, yet her response to both home and waste alters as she lives through the days she writes of. In part, she changes as she draws closer to death, sensing herself as becoming more "waste-like" and in need of Vercueil's aid. Yet the narrative also supplies an important cause in the experiences that push her to reassess the value of home. Four sections of narrative can serve as pointers to her transformation. In the first two brief opening paragraphs, Curren associates Vercueil, whom she views as one of a class of homeless poor wandering the city, with the physical space outside the walls of her house, with those decomposing leaves piling up as litter, a place that is "waste" for being haunted by the memory of children now adult and departed. As homeowner she responds typically, with irritation and an order to leave, but then—as homeowner and indeed liberal—she replaces that moment of dismissal with a qualified generosity. Before long, to justify the help she gives, she sets him to tidying the garden, even though she has had no interest in maintaining it, having let it become "overgrown" (20). That the work is make-work only emphasizes the extent to which her behavior is governed by a conventional awareness of social relations.

The second sequence involves the visit to Guguletu that supplies the episode cited above. There Curren finds houses—and shacks at the adjacent Crossroads—that are no shelter from the pervasive violence encouraged or inflicted by agents of the state. When the sights she faces become too much for her, she says to her companion Mr. Thabane, "I must get home soon." And he replies, "But what of the people who live here? When they want to go home, this is where they must go" (97). The third sequence involves the moment when, shortly afterward, her own house is taken over by the police in a way that reinforces how home, despite earlier assumptions, offers no barrier against the world. When the policewoman tries to bring Curren inside, she calls out "It's not my home anymore!" (157) and wanders down the street to spend her night under the overpass. And the final sequence involves Curren inviting Vercueil into the house and then into her bed (so she can feel the warmth that her cold body craves). She finds that with Vercueil living in the house with her, her bedroom floor is littered with the waste paper he drops, which he picks up when asked to do so but never enough to keep the floor clean (189). No longer is she the one speaking the disdain of the comfortably situated, as when she comments sharply to Vercueil soon after meeting him "You are wasting your life," to which he indicates his opinion with a very material "gob of spit" (8). Nor is Curren the one who, in a more generous moment at the piano, plays Bach, with the thought that somewhere out there in the garden Vercueil the vagrant, object of her charity, might be listening (24). Instead, weakened by pain and medication and lying in her bed, she fusses over him, her "shadow husband": "That is why you should not be so alone. Because I may have to go away entirely" (189, 188).

By narrative's end Curren has lost any illusion that "home" marks a barrier between her and the world, that its walls represent a sheltering distinction between inside and outside, between order and litter, between normalcy and outrage. Her experience progressively undermines a notion that home affords a place well tended and made tidy, by a servant if not by herself, in which walls provide her protection from a "waste" associated with outside, as in her garden, which she talks about as "mess," an incipient "wilderness" (21). If the house as home provides a material context for Curren's shifting subjective responses to the social position bequeathed to her, so too does the landscape within which the house is situated, constructed by the novel in accord with apartheid's political geography. The two-story house with garden in the white city center marks a very different sort of space from the small "matchbox" houses of the black "township" of Guguletu and the shacks in their agglomerations that grow up at its borders. As is typical in the novels preceding it, *Age of Iron* supplies this landscape with a sense of temporal depth that implicates place in colonial history. The history, once again, is suggested, not elaborated, as when Curren makes explicit in confession to Vercueil that "A crime was committed long ago. How long ago? I do not know. But longer ago than 1916, certainly. So long ago that I was born into it. It is part of me, I am part of it" (164). Another example is when she quotes Virgil and Thucydides as authority to her African auditors, thereby both linking Curren to European arrival in Africa and locating the urge to empire well prior to that arrival. Not commented on by Curren but once more implied through her words is how the world of Cape Town resting on this past is marked materially by the power of the state and its technologies, police and military vehicles, and television and radio. Within the political geography both expressing and reproducing the apartheid state (which the novel implies exists as one particular moment in a sweep of imperial history), Curren's house marks a space of privilege and empowerment. Her transformations, the narrative indicates, derive in important ways from her growing awareness of the material and ethical fragility of such middle-class shelters inherited from colonial pasts.

Just as the narrative presents Curren's subjectivity as deriving from a social position given in history, the narrative presents Vercueil as social product. If Vercueil, like Michael K, is to be read as "coloured" (in his South African world), then he serves as a reminder of an additional layer of damage that societies with roots in colonialism have inflicted historically.[8] Associated with people descended from the Cape slave world, Vercueil, as "non-white," lives in this corner of the modern industrial world as one of its expendable people, hired as laborer on a trawler, his wounded hand leaving him as social detritus, his homelessness and alcohol augmenting his sense of not belonging. Interestingly, Curren early calls him "derelict," suggesting abandonment, in a way that echoes her perception of the alley beside her house as abandoned by children, and so the word *derelict* connects Vercueil through the narrator's first paragraph to other words that she applies to the alley: "dead place, waste, without use." The word *derelict* also links him to her vision of the countryside, "the abandoned farmhouses" on "land taken by force, used, despoiled, spoiled, abandoned in its barren late years" (25). And, of course, in this narrative meditation on decay provoked by the onset of a terminal cancer, his "derelict" status links him to her but with this key difference. The sense of being waste that informs Curren's self-awareness has to do with age and the illness that leads to organic breakdown. However painful, this way of being waste can seem natural, generational, linking her to the dead leaves that have piled up in the alley and to the seasonal setting, a Cape winter. Vercueil's being used and abandoned has its root in a colonial past that involved the "ravishing" of land "loved only in the bloomtime of its youth, and therefore, in the verdict of history, not loved enough" (26). Vercueil's status as waste has its origins in the imposition of a specific civilizational form in the land of his birth that, from the start, has marginalized him, just as it has created marginal places—Guguletu and even more so Crossroads—that are marked by "waste," both physical litter and abandoned human potential. In such places, it is the young especially, the narrative suggests, who can suffer both morally and physically as they fight against these imposed conditions.[9]

Whatever was the case with older empires, societies transformed by the capitalism fueling modern colonialism (with an economic logic that prevents full, dignified employment) produce in their marginal places, whether urban or rural, populations of the discarded and derelict. These marginal places, moreover, are themselves marked by litter, with their inhabitants, uprooted from any old economic forms providing their own material order, unable to afford the amenities that support a modern tidiness. Vercueil can be read as figuring such people, and in this context it is worth recalling the origin of his injury in the industrial exploitation of fish populations, which the narrative associates with industrial chicken production (186, 41–44). Industrial capitalism produces classes of people to wring the necks of chickens (or be meatpackers, generally), people of low status, drawn frequently from the previously colonized, who are vulnerable to accident and material insecurity. Moreover, the narrative suggests that in a postapartheid world, these impositions, these deformations of human potential, will more likely rearrange themselves than cease. It is the reliable and stern matriarch Florence who calls Vercueil "rubbish," "good for nothing," a "rubbish person" (47, 59): the civilization set in motion during the colonial era and inspired by histories of empire will continue to create human waste and then view it with contempt or treat it with violence.

What fascinates about *Age of Iron* is how the idea of "waste," with its cluster of related ideas, invites a reading that draws together postcolonial and ecocritical interpretations. Apartheid's political geography has its roots in colonial spatial control, which also (if the 1913 Natives Land Act is recalled) amounts to a control of human populations with economic intent. The creation of marginal lands in a modern industrializing colonial society is the creation of marginal peoples—both settler descendants pushed for economic and cultural reasons away from colonial centers and indigenous people excluded from them, who become people of color. In this regard, *Age of Iron* encourages rereading Frantz Fanon's essay "Concerning Violence," with its linkages of violence and a colonial political geography mapping "two zones," "the settlers' town" and the "native towns . . . wallowing in the mire," with their scorn for the colonizer's "civilized" values.[10] Perhaps the most illuminating detail in Fanon's text to read beside *Age of Iron* is the following: "All that the native has seen in his country is that they [the colonialist bourgeoisie] can freely arrest him, beat him, starve him: and no professor of ethics, no priest has ever come to be beaten in his place, nor to share their bread with him" (44). Curren may not step in to share the physical beatings meted out by apartheid police, but she does jettison much of the constitutive spirit of the "colonialist bourgeoisie." Yet in choosing Curren as focalizer, Coetzee stages an important perspectival difference: the novel explores the apartheid landscape from within the mind of a "settler," one who benefits from the order that passes and thus does not participate in the hunger for revolution, but also one who despises the order that benefits her and does not elegize its passing. This conflict-ridden subjectivity affords Coetzee the opportunity to highlight (without implying a potential for revolutionary change) what modern life tries hard to sweep and flush—like actual waste—out of mind: the tendency within modernity to deploy technologies to create, categorize, and police distinct places, thereby dominating space and subordinating populations, while creating an undertow of destructiveness in which what can be used is taken with little moral regard and what has been used up, or has no use, is discarded as refuse.

Reading the novel beside its contemporary South African environmental and political discourses reveals the affinity of its vision of waste, not with environmentalism, which still in the 1980s articulated a relation to the material world—and to local people—that reproduced colonial discourse and practice, but with an antiapartheid discourse that focused on this society's producing people as "surplus" and "discarded."[11] Since the transition to democracy, there may have been changes to South Africa's environmental culture that address the quality of all people's lives, including the poor, but within this culture's various strands, from environmental justice activism to wildlife preservation, there remain inevitable sources of tension (mirroring those in the global arena).[12] Not surprisingly, *Age of Iron*, in its suspicion of modernity, still exposes limits to mainstream forms of environmentalism, including those that take up the issue of waste. Bypassing anxieties over excessive consumption (Curren's car marks her as very much not a conspicuous consumer), the novel reinvigorates in postcolonial mode earlier meanings of "waste" (ranging over "unproductive," "useless," "left over" and "discarded") and thereby exposes a kind of meaning-making to which the prosperous world, wanting to care for nature while maintaining its comforts, can appear blind.[13] Curren as focalizer allows Coetzee to develop perceptions of the modern world that recall those that K's outsider status affords him, a sense of modernity as involving not so much a disciplining and containment of the unruly (an activity of which *Michael K*'s "camps" are emblematic) as a dividing of life into the useful and the useless and the discarding as waste both what it finds useless at the outset and what it "uses up."

With this perspective on modernity, it is not surprising that *Age of Iron* can continue to subvert a mainstream discourse of waste. Waste has indeed become a concern in corporate and government circles accompanying worry about sustainability in South Africa's industrial economy. And in ways that reflect tendencies among Northern societies with municipal recycling programs, this environmental discourse has adopted the term *zero waste*. Gaining prominence in the North in the late 1990s, zero waste develops the sense of waste as excess, as abundance discarded, that derived from the North responding to the widespread consumerism that followed World War II and articulates mounting anxieties over First World prosperity appearing to be threatened by natural resources' finitude.[14] As early as 2001 the South African government, at its "first National Waste Summit," took as its goal to "Reduce waste generation and disposal by 50% and 25% respectively by 2012 and develop a plan for ZERO WASTE by 2022." Zero waste as a goal has since found its way into planning by local municipalities, as evidenced by Victor Munnik's study of a process that involved civil society in Mogale City, Gauteng, in the preparation of an Integrated Waste Management Plan. Zero waste, moreover, has a South African NGO, IZWA, devoted to its promotion and has been advanced on prime-time television in the scripts and supporting advertisements for *Hybrid Living*, a television magazine show sponsored by Toyota, which has a supporting Web site (with the banner "Creating Eco-lifestyles").[15]

What interests about this literature on zero waste (and waste generally in Northern countries but echoed in South Africa) is a tendency to ignore social divisions, to ignore the fact that an industrial economy produces classes of people along with its consumer goods. Nor does it have much to say about the global economy as rooted in colonial history, about the traffic across oceans of both goods and garbage, and of the outsourcing of both production and garbage clean-up, so there is no sense either that the industrial production of classes of people may define social worlds not only locally but also on a global scale.[16] In this regard, zero waste discourse seems to echo much wilderness discourse in its historical and sociological thinness. This modern world does, of course, need to be concerned with resource use, with the consequences for both humans and natural systems of ever-expanding global trade, and so it does need to absorb the vision inspiring the zero waste movement, with industrial economies mimicking as much as possible the closed loops of natural systems. Ecology—together with environmental history—has this to teach modern societies. But simultaneously we who are beneficiaries of modernity need to address the problems that modernizing has bequeathed to societies, from the collapse of traditional agricultural communities to the growth of urban slums to the anomie attending meaningless and ill-paid labor, an accumulated inheritance that contradicts modernity's emancipatory promise. One need only recall the size of the planet's slum populations to sense the grotesque proportions of this failure.[17]

In its postcolonial suspicion of modernity, Age of Iron bears more relation to an African writing about waste that does not echo the concerns of the prosperous North. Such writing concentrates for obvious reasons on waste as refuse rather than as a problem deriving from overconsumption and focuses on the difficulty in urban environments of waste management. Nimbe Adedipe has said of the zero waste principle that it is a "desideratum that African countries must adopt," but the remainder of his essay concerns the practical problems facing effective waste treatment where economies are insufficiently robust and where, in particular, "the rapid rate of uncontrolled and unplanned urbanization" results in an "inability to match generation rates with collection and disposal rates."[18] Most studies supplement a focus on scale with details regarding the role of economics and cultural politics: it is the poorest, most marginalized of city dwellers who are condemned to live among refuse and, where refuse-collection services exist, who suffer the low wages that such work brings.[19] Garth Myers, in his rich study of waste disposal as an aspect of urban government in three African cities—Dar es Salaam, Zanzibar, and Lusaka—traces how colonial cultural practices (including city planning) and current neoliberal economic forces (and accompanying ideologies) intersect to create urban spaces in which livelihoods for a majority are precarious and life is led in settings marked by accumulating refuse.[20] Of course, waste can offer opportunity to the poor. In the South, where municipal services are thin and poverty levels are high, "subsistence waste-picking" is widespread.[21] Among shanty dwellers, discarded plastic, cardboard, and paper can be useful for providing material for insulating and decorating homes. But not all waste is benign: for the poorest city dwellers, excrement is a prominent environmental hazard. Recent years have seen news reports of flying toilets whereby plastic bags (commercial waste) is being recycled to solve a problem of the body's waste in places where organized sanitation is inadequate or lacking.[22]

Such empirical studies substantiate Age of Iron's shaping in its realist dimension of the nature-human relation, in which some people and some places become modernity's waste products outside its regime of tidy regularity. Perhaps it is South African academics developing an environmental justice perspective that, in their combination of analysis and condemnation, have developed lines of thinking close in spirit to this aspect of the novel. For David Hallowes and Mark Butler, place is constructed socially, and places of poverty suffer simultaneously from environmental degradation and social conditions that hinder full human development. They note how the stress of life in such deeply impoverished environments induces "psychosocial disorders and diseases" that bear fruit in destructive conflict.[23] Hallowes and Butler comment on waste while pointing to the particular form that apartheid policies gave to South African cities. It is in the "peri-urban settlements" that people suffer the accumulation of "human waste and garbage," while in settlements near industry people suffer from discarded toxic waste (69, 71). Writing with a historical perspective, Hallowes and Butler link modernity and colonial processes, grasping modernity's implication in South Africa's political geography, a geography that continues to bear the traces of modernity's uneven development, and they oppose a mainstream discourse of sustainable development in which ecologically sustainable economic progress, built on current social arrangements, brings into the mainstream economy ever more people, in favor of discourse that recognizes that current socioeconomic arrangements inevitably produce poverty and ecological damage (59).

Yet *Age of Iron,* of course, distances itself from scripts inspiring activism. While the narrative develops its vision of the harms inflicted on this part of Africa, the narrator whom Coetzee chooses is a retired classics scholar who, moreover, while she weakens into painful death is revealed to be ineffectual in her interventions and flustered, uncertain, and excessive in her musings. Through Curren, the novel develops a way of naming the world that is replete with the sort of subjective complexity that managerial and activist discourses shun for blocking successful accomplishment. It is in this complexity, which includes a postmodern alertness to subjectivity's linguistic dimension, that *Age of Iron* stands apart from what emerges as South Africa's environmental justice writing, despite their shared focus on harms caused by colonial modernity. Coetzee provides a measure of this distance by having Curren, in conversation with Vercueil, articulate a yearning for a living female-focused world outside the course of patriarchal history. In a passage that recalls a moment when K, remembering the rules at Huis Norenius, thinks, "They were my father, and my mother is buried and not yet risen" (*Life & Times*, 105), Curren suggests the richly felt and imagined connection to a unified world that current conditions have taken from her: "A desire, perhaps the deepest desire I am capable of, would have flowed from me toward that one spot of earth, guiding me. *This is my mother,* I would have said, kneeling there. *This is what gives life to me.* Holy ground, not as a grave but as a place of resurrection is holy:

resurrection eternal out of the earth" (*Age of Iron,* 121). If Curren is read with K's language in mind, she (or Coetzee through Curren) seems here to push thought once again toward a version of reality that is painfully nonexistent, countered by an actuality developed within the sweep of a male-dominated time in which "war" has been "the father of all things," in which the world has arrived at an "age of iron."[24]

The novel nonetheless is not defeatist. As noted above, Curren begins the narrative as one who unwittingly reinforces apartheid's political geography through her participation in it as homeowner, finding solace in its shelter. The argument could be made that she never moves beyond associating organic decomposition with social order's breakdown and associating rotting leaves with litter. Yet she does pull the decaying into her heart and make her peace with rot and untidiness, her experience having suggested how her social world compromises an appearance of healthy order.[25] Thus, she ends the narrative having detached herself, in attitude and behavior at least, from the privileges that homeownership supplies to the point of finding both intimacy with Vercueil and sympathy, if not identification, with revolutionary violence (165). Curren, in her changing response to both Vercueil and "John," develops a complex understanding of love as quite different from the love she remembers for her daughter as child. Curren learns to question the love that is easy, to ponder instead whether loving what initially is grasped as unlovable may not be a truer kind of love than the love she remembers for her daughter (136, 5). Although Curren does not note this—her thinking remains imprecise, emotionally driven—the unlovable whom she considers do not include the obviously unlovable policemen and apartheid politicians. It is those who have been damaged and discarded by history whom she reaches toward, not those who have inflicted the hurt, and this emotional response accords with her intuition that waste has an origin in an inadequacy of love (25–26).

This intuition, because the decay that Curren senses around her is pervasive, becomes in the developing narrative an illuminating comment on modernity itself, not simply on apartheid. The language, as occurs so often in this novel, reaches beyond its setting to indicate a more general modern failure: "our hearts . . . long for these sedate homes of ours to tremble . . . with angelic chanting." This will not happen, however, "not in the suburbs. The suburbs, deserted by angels" (14). But these are words she speaks early in her narrative, and the kind of love she discovers through her spiritual journey takes her from such talk of angels. The world remains as empty of the numinous, but she finds a way—even as and in part because she sickens into death—to enact a spiritual healing that makes such talk pointless. Curren begins to be aware of love as an experience that brings the privileged from their confining cultural and social inheritances out into actual worlds where they appear as oppressive or ignorant, and through this journey she finds a way to bring her subjectivity closer to alignment with actuality. Yet while this subjective journeying figures a healing of the divide between Curren's subjectivity and her historical, material circumstances, these circumstances offer a position from which to comment on its value. Her journey matters, but the historical forces she is up against dwarf the journey.

To the extent that this writing from "the jaws of death" bears "truth," Curren's narrative suggests that the damage inflicted by modernity includes an ethical damage to the privileged (us) and that any recovery will entail a painful and most likely inconclusive reaching against this tide of forces from the past toward what they have made Other. The novel speaks to the crucial importance of keeping a long view of modernity's (and empire's) interaction with the material world and the consequent creation of poverty, of marginalized people, and of waste lands. In this attention to modernity's past, the novel both reaches beyond the particularities of one woman's life in 1986 in a Cape Town torn by apartheid-generated violence and bypasses any obvious politics. The kind of political awareness that the novel stimulates to qualify its exploration of ethics has less to do with issues of who controls the state than with issues surrounding homeownership in such a society. Politics not only intrudes into the home (such as in the initial distinction between Vercueil the vagrant and Florence the trusted servant) but also *begins with* the home, because with property ownership comes a subjectivity that separates the homeowner from those whose access to home is either tenuous or lacking. Curren, as noted, reveals a class-based tendency to associate usefulness and productivity with material fruitfulness and social order and then to associate their contraries with material decay and social disorder, even while she loses her distaste for these. Thus, the political awareness elaborated has more to do with attention to the power relations exposed by modernity's political geography and history than with questions, however urgent, of which group of people might own land or with parties and policies, whether liberal or conservative or authoritarian. Absent from the novel—except in Curren's wish for it—is any social order, materially fruitful, not shaped by this deep-rooted "age of iron" (50).

This kind of radically suspicious political awareness complicates the novel's insights into a suitable ethics, even as the novel proposes the primacy of ethics over political action. The novel, I believe, recommends that its readers engage in a similar complicating maneuver, asserting the primacy of the ethical over a dogmatic politics—the dogma that politics is forever in danger of sinking into—while simultaneously grasping the limitations placed on that project by the social positions that we as comfortable readers occupy. All societies make waste, but societies currently dominant, rooted in European global expansion and formed within a regime of global industrial capitalism, make waste in distinctive ways. *Age of Iron* suggests something of the extent of this society's power to devastate both people and place, and so the novel can, I believe, open a perspective on the complexity of environmental work needed on the African continent, a work needed to counter current modernity's continuous damage. Such crossing of history's boundaries as Curren engages in, necessitating first a relinquishment of material and intellectual privilege and then an enlarged, informed expression of love, one that grasps its relation to necessary and not always appreciated work, could open an understanding for its middle-class readers of an environmentalism very different from a mainstream green preservation of wild nature (while ignoring modernity's production of suffering from poverty) or concern with ensuring that natural resources remain for sustainable maintenance of the social status quo. Yet the novel (obviously) does not offer Curren's transformation as any sort of solution. For those of us engaged in postcolonial ecocriticism, Curren's narrative suggests that no easy solutions exist to repair what modernity has damaged and that no action in the field of such damage carries value unless it is predicated on a renunciation of all that is comforting—assumptions, emotions, or experience—unless, in other words, it is predicated on "biting the dust" (195).[26]

In ways that are characteristic of Coetzee's novels, *Age of Iron*'s power stems from its quest for human value while denying credibility to social movements for general emancipation. The novel thus works as a corrective to the self-congratulation available to the comfortable and, more generally, to a reformist (or revolutionary) optimism rooted in the European Enlightenment.[27] Setting itself against modernity's obvious and powerfully attractive promise, a promise that globally draws the desires and hopes of so many (and in 1986 was available to the privileged in South Africa), *Age of Iron* articulates crucial dimensions of what modern success hides and that even if observed appear too dissonant to integrate. Yet the narrative's disillusionment with modernity—more so an active disillusioning—speaks to a refusal to let modernity be distracted from how it fails humanity. In Curren's journey, *Age of Iron* offers a glimpse into a humanity that is possible through such disillusionment, and while this valued humanity is not one grounded in a socially productive solidarity, let alone a practical politics, it signals minimal conditions that such solidarity might need to meet. More-over, by presenting its observations through a mind at "the jaws of death," the novel suggests that it is commitment to life, experienced as commit-ment to modern worlds, that keeps hidden, or fleetingly noted, the cost that modern worlds exact. Beyond such plausibility in its realist dimen-sion, the novel, through its postmodern focus on textuality, tangles readers in its abstract questioning. While we may not live in that place and time, we nonetheless are always meaning makers, sharing the modern discourses that offer success and taking shelter in the homes they enable, and our at-tempts to make meaning—as in this chapter—might, for shanty dwellers, amount simply to "shit." If there is "truth" to *Age of Iron*'s narrative, so lay-ered in irony, then it includes the (postmodern) thought that only outside such shelters does grappling with reality begin.

Notes

For supporting the research, I thank Transylvania University's Kenan Fund for faculty and student enrichment

1. See Rob Nixon, "Environmentalism and Postcolonialism," in *Postcolonial Studies and Beyond,* edited by Ania Loomba, Suvir Kaul, Matti Bunzl, Antoinette Burton, and Jed Esty, 233–51 (Durham, NC: Duke University Press, 2005). Intro-ductory overviews of work done to link environmentalism and postcolonialism appeared soon thereafter; see Graham Huggan and Helen Tiffin, "Green Post-colonialism," *Interventions* 93, no. 1 (2007): 1–11, and Anthony Vital and Hans-Georg Erney, eds. *Postcolonial Studies and Ecocriticism,* Special issue of *Journal of Commonwealth and Postcolonial Studies* 13, no. 2, and 14, no. 1 (2006–7): 3–13. For a recent book-length study, see Graham Huggan and Helen Tiffin, *Postcolonial Ecocriticism: Literature, Animals, Environment* (London: Routledge, 2010).

2. See Nixon (in this volume) for his analysis of such violence and its struc-tured invisibility; see also Caminero-Santangelo (in this volume) for analysis of hiddenness in terms of a political/environmental unconscious. Caminero-Santangelo's strategies for exposing this "unconscious" promise a fruitful way to grasp literary texts—and daily life—in terms of ideology. *Age of Iron,* I argue in this chapter, presents its own study of such hiddenness. Reading the novel with a materialist focus on discourse—reading its literary discourse beside relevant nonliterary discourse while mindful of social and cultural contexts—highlights both the scope of the novel's vision and the way that the novel ex-plores an ethical response to what it opens to view.

3. J. M. Coetzee, *Dusklands* (New York: Penguin, 1983), *Life & Times of Mi-chael K* (New York: Viking, 1984 [1983]), and *Age of Iron* (New York: Vintage

Books, 1990). For ecocritical readings of *Michael K*, see Dominic Head, "The (Im)Possibility of Ecocriticism," in *Writing the Environment: Ecocriticism and Literature*, edited by Richard Kerridge and Neil Sammells (New York: Zed Books, 1998), 27–39, and Anthony Vital, "Toward an African Ecocriticism: Postcolonialism, Ecology and *Life & Times of Michael K*," *Research in African Literatures* 39, no. 1 (2008): 87–106 (corrected version, http://www.postcolonial-ecology.net). For comment on Coetzee's development of postcolonial narrative forms, see Vital, "Toward an African Ecocriticism," 99–100. In connecting present worlds to colonial pasts, thereby elucidating the global in the local, these narratives draw on a structuralism derived from linguistics rather than on a historical materialism. Through this turn toward textuality, Coetzee's writing reflects a postcolonial suspicion of nationalism and liberation discourse, one that articulates awareness that modernity develops inescapable systems of domination. The fiction, while emancipatory in intent, is sharply at odds with the socioeconomic analyses of the traditional Left, analyses that seek within modern structures for a liberation of human potential.

4. Derek Attridge, *J. M. Coetzee and the Ethics of Reading: Literature in the Event* (Chicago: University of Chicago Press, 2005), provides valuable comment on *Age of Iron*'s narrative: "To read *Age of Iron* is to read, or overread, a strange kind of letter, written in 1986 by a dying woman in Cape Town to her married daughter in the United States. This, at least, is the fictional contract we enter into, though we are given little in the way of realistic reinforcement that might enable us to imagine the words issuing from a pen onto a sheet of paper" (91). Moreover, if Curren in the last sentences records the moment of her death, as is hinted, then the writing closes with reinforcing its artifice. Attridge examines how the novel deploys this formal complexity to pose ethical questions suited to a postcolonial world marked by state violence. For a similar focus, see two essays by Michiel Heyns: "An Ethical Universal in the Postcolonial Novel—'A Certain Simple Respect,'" in *Thresholds of Western Culture: Identity, Postcoloniality, Transnationalism*, edited by John Burt Foster Jr. and Wayne Jeffrey Foreman, 103–13 (New York: Continuum, 2002), and "Houseless Poverty in the House of Fiction: Vagrancy and Genre in Two Novels by J. M. Coetzee," *Current Writing: Text and Reception in Southern Africa* 11, no. 1 (1999): 20–35. For a reading that focuses on the novel's epistemological dimension, see Dominic Head, *J. M. Coetzee* (New York: Cambridge University Press, 1997). For readings that explore the topics of love and the fragility of literature's insights against the crude violence of history, see Michael Marais, "Places of Pigs: The Tension between Implication and Transcendence in J. M. Coetzee's *Age of Iron* and *The Master of Petersburg*," *Journal of Commonwealth Literature* 31, no. 1 (1996): 83–95, and Michael Marais, "'Who Clipped the Hollyhocks?': J. M. Coetzee's *Age of Iron* and the Politics of

Representation," *English in Africa* 20, no. 2 (1993): 1–24. For readings of Curren's ethical journey that focus on intertextual reference (to Virgil's *Aeneid* and Dante's *Divina Commedia*), see Lars Engle, "Western Classics in the South African State of Emergency: Gordimer's *My Son's Story* and Coetzee's *Age of Iron*," in *Thresholds of Western Culture*, 114–30, and David E. Hoegberg, "'Where Is Hope?': Coetzee's Rewriting of Dante in *Age of Iron*," *English in Africa* 25, no. 16 (1998): 27–42. My reference to history, noting how materiality provides implicit commentary on Curren's thought, is different from that of Graham Huggan, "Evolution and Entropy in J. M. Coetzee's *Age of Iron*," in *Critical Perspectives on J. M. Coetzee*, edited by Graham Huggan and Stephen Watson (London and New York: Macmillan/St. Martin's, 1996), which unpacks how Curren makes sense of her life by infusing thinking about history with "mythmaking" (203). My reading of the novel for its relation to patriarchal modernity differs in approach from that of Laura Wright, "Displacing the Voice: South African Feminism and J. M. Coetzee's Female Narrators," *African Studies* 67, no. 1 (2008): 11–32, which reads Coetzee's choice of a female narrator as illustrating how the ambiguous position of South African white women in a colonialist patriarchy "silence[s] a . . . feminist agenda" (12).

5. Cf. Coetzee's observation in *Doubling the Point: Essays and Interviews*, edited by David Attwell (Cambridge: Harvard University Press, 1992), that on one level the novel "stages" a "contest" between Curren's voice and "voices of history" that would "deride" her for the private nature of her struggle, for representing the classics, "from long ago and far away" (250). Note too how he encourages perception of the biological limit to her words' value, raising the issue of "whether Elizabeth has the right to speak," whether she has any authority as one speaking "from the jaws of death" (340). In the fragility of her opposition to the status quo, Curren prefigures Costello; see Anthony Vital, "Situating Ecology in Recent South African Fiction: J. M. Coetzee's *The Lives of Animals* and Zakes Mda's *The Heart of Redness*," *Journal of Southern African Studies* 31, no. 2 (2005): 302.

6. See Hoegberg, "'Where Is Hope?,'" and Engle, "Western Classics in the South African State of Emergency."

7. See Head, *J. M. Coetzee*, on her "increasing sense of her own insignificance" and on how she "comes to terms with the irrelevance of her ideas" (140, 130).

8. The narrative does not explicitly classify Vercueil's "race," a silence reinforcing Curren's—and Coetzee's—refusal to grant validity to the race thinking elaborated by colonials and insisted on by the apartheid regime. Yet as in *Michael K*, the text supplies suggestions that Vercueil, like K (unlabeled except by a state official), is meant to be understood as "coloured": his laboring on a trawler and his treatment by John and Bheki, who include him in a "we"

opposed to a "they," which they figure as white (45). Vercueil, representing the *vagrant* as Other, needs to be read in relation to both *Michael K* and Coetzee's earlier exploration of idleness in the "Discourse of the Cape," first published 1982 and reprinted, while Coetzee was composing *Age of Iron,* as the first chapter in his *White Writing: On the Culture of Letters in South Africa* (New Haven, CT: Yale University Press, 1988).

9. Hoegberg provides a valuable account of the events in Crossroads to which Coetzee's narrative refers and also points to how the young Comrades could be seen as ethically compromised ("'Where is Hope?,'" 32–36). For discussion, drawn from sociology and anthropology, of the contemporary production of humans as waste, see Zygmunt Bauman, *Wasted Lives: Modernity and Its Outcasts* (Cambridge, UK: Polity, 2004), and João Guilherme Biehl, *Vita: Life in a Zone of Social Abandonment* (Berkeley: University of California Press, 2005), a study of life in Porto Alegre, Brazil, confronting "how to restore context and meaning to the lived experience of abandonment" (23). As I suggest later, Fanon's writing illuminates such topics well for a reading of *Age of Iron.*

10. Frantz Fanon, *The Wretched of the Earth,* translated by Constance Farrington (New York: Grove, 1963 [1961]), 88, 39, 43.

11. For an account of South Africa's environmental culture, see Farieda Khan, "The Roots of Environmental Racism and the Rise of Environmental Justice in the 1990s," in *Environmental Justice in South Africa,* edited by David A. McDonald (Athens: Ohio University Press, 2002), 15–48. For the social production of people as "waste," see the activist scholarship of Laurine Platzky and Cherryl Walker, *The Surplus People: Forced Removals in South Africa* (Johannesburg: Ravan, 1985), and Cosmas Desmond, *The Discarded People: An Account of African Resettlement in South Africa* (Harmondsworth, UK: Penguin, 1970). The apartheid government's ponderous reference to the wives and children of African urban laborers as "superfluous appendages" admitted into policy such transformation of people into waste. For an account of Nadine Gordimer's exploration in *The Conservationist* (London: Penguin, 1972) of apartheid's production of people as waste, see Rita Barnard's chapter "Of Trespassers and Trash" in her valuable study, *Apartheid and Beyond: South African Writers and the Politics of Place* (New York: Oxford University Press, 2007). Barnard reads for Gordimer's attention to the cultural-political significance of place; my intent in this chapter (on Coetzee's very different novel) is to read place with an eco-social concern suited to African conditions.

12. See Jacklyn Cock and David Fig, "The Impact of Globalisation on Environmental Politics in South Africa, 1990–2002," *African Sociological Review* 5, no. 2 (2001): 15–35, CODESRIA.Org, http://www.ukzn.ac.za/ccs/default. asp?6,20,10,1809.

13. Entering English as translation of its Latin root, "vastare," the term *waste*, in its early usage, reveals how its reference bridges nature and society, conveying ideas of emptiness and desolation, of being desert, lifeless, and unproductive, early meanings that link waste to the term *devastate* and to the phrase "to lay waste." Early usage, in other words, carries the anxiety associated with an agrarian society's need for productive cropland. By the end of the fifteenth century the word *waste* is used to intend, in addition, ideas of both useless expenditure (as in waste of money) and refuse matter (as in waste removal). The anxiety attached to material scarcity in an increasingly commercial society can be heard in the proverb "waste not, want not," first recorded in 1772, with its call for wise husbandry of resources. See Wolfgang Mieder, Stewart A. Kingsbury, and Kelsie B. Harder, eds., *A Dictionary of American Proverbs* (New York: Oxford University Press, 2000). Such usage makes sense in a modernizing Britain, and in this context the harshness of punishment meted out in Elizabethan England to "rogues, vagabonds and sturdy beggars" (as they are commonly referred to) appears related: avoiding productive hard work wastes a community's potential for increased wealth. This usage can be linked to Coetzee's elaboration (through Curren) of Vercueil as well as to Coetzee's exploration of idleness in *White Writing*.

14. This concern over finite resources finds expression in the widely read report by Donella H. Meadows et al., *The Limits to Growth: A Report for the Club of Rome's Project on the Predicament of Mankind* (New York: Universe Books, 1972). The governing idea behind zero waste is simple, as the Zero Waste Alliance's Web site (Zero Waste Alliance, a Program of the International Sustainable Development Foundation, http://www.zerowaste.org/case.htm) suggests, and the attached anxiety is in the imperative: "We recognize that nature is cyclical and has no waste. Our industrial pathways must follow this design set by nature if we are to become sustainable." In other words, throughout society, especially within its industrial organizations, "each material must be used as efficiently as possible and must be chosen so that it may either return safely to a cycle within the environment or remain viable in the industrial cycle." Zero waste in this way becomes what the Web site calls "a key to our grandchildren's future." NGOs have played an important role in pressuring government and industry to make policy changes; in 2002, Greenpeace UK published a rigorous and extended policy document by Robin Murray, *Zero Waste* (London: Greenpeace Environmental Trust, 2002), with the following press release: "Britain Could Be a Rubbish-Free Society Says Ground-Breaking Study," March 18, 2002, Greenpeace UK, http://www.greenpeace.org.uk/media/press-releases/britain-could-be-a-rubbish-free-society-says-ground-breaking-study.

15. *Hybrid Living,* "an eco-friendly SA show about how to live a modern lifestyle while going green at the same time," links its sponsor, Toyota, with zero

waste through the corporation's "Zero Waste to Landfill" commitment; see *TVSA Forums Premiere: Hybrid Living (SABC3)*, TVSA, http://www.tvsa.co.za/ forum/calendar.php?do=getinfo&e=2729&day=2008-4-30&c=1, and The Toyota Corporation's *Environmental and Social Report 2004*, Toyota South Africa, www.toyota.co.za/toyotaworld/enviro_report_2003_4.pdf. For the "First National Waste Summit," see Department of Environmental Affairs and Tourism, South African Government, "The Polokwane Declaration on Waste Management," The First National Waste Summit, Hosted by the Department of Environmental Affairs and Tourism, Pietersburg, September 26–28, 2001, http:// www.environment.gov.za/ProjProg/WasteMgmt/Polokwane_declare.htm. For the work in Mogale City, see Victor Munnik, *Learning with Mogale City: Civil Society Participation in Developing an Integrated Waste Management Plan* (Johannesburg: Group for Environmental Monitoring, 2004). For the IZWA, see Muna Lakhani, "A Zero Waste Local Economy: Myth, Mystery or Magic?," Prepared for the ICLEI Conference, Cape Town, 2006, Institute for Zero Waste in Africa–IZWA, Chintan Environmental Research and Action Group, Information from Other Organizations, June 3, 2008 www.chintan-india.org/others/ zero_waste_local_economy_iclei_booklet.pdf.

16. Reports continue to appear about toxic waste from the North being shipped to countries of the South. See, for example, Paul Ogbuokiri and Stan Okenwa, "Toxic Ship Berths in Lagos," June 4, 2010, AllAfrica Global Media, http://allafrica.com/stories/201006040343.html.

17. See Mike Davis, *Planet of Slums* (New York: Verso, 2006), 22–26.

18. Nimbe O. Adedipe, "The Challenge of Urban Solid Waste Management in Africa," in *Rebirth of Science in Africa: A Shared Vision for Life and Environmental Sciences*, Contributions to the African Renais-Science Conference Held at the Durban Botanic Gardens Visitor's Complex, March 25–29, 2002, edited by Himansu Baijnath and Yashica Singh (Hatfield, South Africa: Umdaus, 2002), 175, 179.

19. See Mark Swilling and David Hutt, "Johannesburg, South Africa," in *Managing the Monster: Urban Waste and Governance in Africa*, edited by Adepoju G. Onibokun (Ottawa, Ontario: International Development Research Centre, 2000), 173–225; see also Msokoli Qotole, Mthetho Xali, and Franco Barchiesi, *The Commercialisation of Waste Management in South Africa*, Research Series #3, 2001, Queens University, Canada, Municipal Services Project, http:// www.queensu.ca/msp/pages/Project_Publications/Series/3.htm, and Melanie Samson, *Dumping on Women: Gender and Privatisation of Waste Management* (Woodstock, South Africa: Municipal Services Project and South African Municipal Workers' Union, 2003).

20. Garth Andrew Myers, *Disposable Cities: Garbage, Governance and Sustainable Development in Urban Africa,* Re-Materialising Cultural Geography Series (Aldershot, UK: Ashgate, 2005).

21. See Farieda Khan, *Waste Picking for Survival: A Report on the Waste Pickers of Frankdale Informal Settlement* (Rondebosch, South Africa: Environmental Advisory Unit, University of Cape Town, 1996). For "rag-pickers" in India, see Jeremy Kuper, "Final Collection," *Guardian,* August 5, 2006, http://www.guardian.co.uk/money/2006/aug/05/careers.work4/print.

22. For recycling among the poorest, see Michael Wines, "Shantytown Dwellers in South Africa Protest Sluggish Pace of Change," *New York Times,* December 25, 2005, http://www.nytimes.com/2005/12/25/international/africa/25durban.html. For the sanitary hazards facing slum dwellers, see Davis, *Planet of Slums,* 137–42; Joyce Mulama, "Menace of the Flying Toilets," Mail & Guardian Online, http://www.mg.co.za/printformat/single/2006-10-26-menace-of-the-flying-toilets; and Mark Whittaker, "Why Uganda Hates the Plastic Bag," BBC News, June 30, 2007, http://news.bbc.co.uk/1/hi/programmes/from_our_own_correspondent/6253564.stm. The United Nations Development Programme notes that "only about 1 person in 3 in Sub-Saharan Africa and South Asia" has access to modern sanitation, and that picture is worse in reality because many modern sanitation systems in urban areas are overstressed and counterproductive; see *Summary: Human Development Report 2006; Beyond Scarcity: Power, Poverty and the Global Water Crisis,* United Nations Development Programme, http://hdr.undp.org/en/media/HDR2006_English_Summary.pdf, 23.

23. David Hallowes and Mark Butler, "Power, Poverty, and Marginalized Environments: A Conceptual Framework," in *Environmental Justice in South Africa,* edited by David A. McDonald (Athens: Ohio University Press, 2002), 73.

24. David Hughes (in this volume) offers a rich examination of settler mystification of land as a means to avoid engaging responsibly with a "native" population; the radical awareness that Coetzee writes into Curren's narrative suggests that her imagined connection with her mother's grave cannot be read in such terms but supplies an authentic (if hopeless) response to a history of settlement.

25. Hoegberg notes that Curren has a tendency when confronted with difficulty to "retreat" toward (seek shelter in) text that her vocation has given her to find meaningful ("'Where is Hope?,'" 27). In this context it is important to stress Curren's growing awareness that she has not been "awake," that she has shared a mode of existence with those "bee grubs . . . drenched in honey" that she scorns early in the narrative (*Age of Iron,* 7). Securing islands of comfort

in worlds where violence is prevalent (whether in South Africa or in North America, as she appears to understand her daughter's case) can isolate, the narrative suggests, and deaden intelligence.

26. Just as events lead Curren to question the "easy" love she remembers for her girl-child, they lead her beyond an "easy" aesthetic response to nature (18). Although *Age of Iron* offers no programmatic thinking, no reflection on the steady, communal, ameliorative work so needed against diminished life chances, the novel could prepare for such clear-sighted work, the sort outlined by Nixon in this volume. Curren, of course, is of European settler, not African, ancestry. Yet while ethnicity would produce subjective differences in the experience of reaching across history's boundaries toward damaged life, the journeys, from whatever starting point, would seem to carry the same moral weight. For a related expression of care, see Ben Okri's writing of modernity's disruption of Africa's traditional foodways, its damaging creation of places of deprivation, as explored by Jonathan Highfield (in this volume).

27. In readings of *Lives of Animals* and *Michael K,* I have commented on the value and limitation of Coetzee's commitment to textuality for an African ecocriticism while suggesting, in the case of *Michael K,* an understanding of this commitment in its historical moment ("Situating Ecology in Recent South African Fiction," 303–6; "Toward an African Ecocriticism," 96, 99–102). *Age of Iron* embodies a similar culturalism, with materiality subject to text and text motivated by a character who is a solitary with little ability to alter circumstances in socially meaningful ways. While valuable for highlighting, through its postcolonial narrative, the need to hold discourse as formative of the nature-human relation, *Age of Iron,* as with the others, skirts elaborating images of the ordinary work (necessarily social) needed to counter the damage in postcolonial worlds. Yet it is through such refusal that the narratives achieve their distinctive critiques of modernity.

Bibliography

Adedipe, Nimbe O. "The Challenge of Urban Solid Waste Management in Africa." In *Rebirth of Science in Africa: A Shared Vision for Life and Environmental Sciences.* Contributions to the African Renais-Science Conference Held at the Durban Botanic Gardens Visitor's Complex, March 25–29, 2002, edited by Himansu Baijnath and Yashica Singh, 175–92. Hatfield, South Africa: Umdaus, 2002.

Attridge, Derek. *J. M. Coetzee and the Ethics of Reading: Literature in the Event.* Chicago: University of Chicago Press, 2005.

Attwell, David. "'Dialogue' and 'Fulfilment' in J. M. Coetzee's *Age of Iron*." In *Writing South Africa: Literature, Apartheid, and Democracy, 1970–1995*, edited by Derek Attridge and Rosemary Jolly, 57–74. Cambridge: Cambridge University Press, 1998.

Barnard, Rita. *Apartheid and Beyond: South African Writers and the Politics of Place*. New York: Oxford University Press, 2007.

Bauman, Zygmunt. *Wasted Lives: Modernity and Its Outcasts*. Cambridge, UK: Polity, 2004.

Biehl, João Guilherme. *Vita: Life in a Zone of Social Abandonment*. Berkeley: University of California Press, 2005.

"Britain Could Be a Rubbish-Free Society Says Ground-Breaking Study." Press Release, March 18, 2002. Greenpeace UK, http://www.greenpeace.org.uk/media/press-releases/britain-could-be-a-rubbish-free-society-says-ground-breaking-study (accessed July 4, 2007).

Cock, Jacklyn, and David Fig. "The Impact of Globalisation on Environmental Politics in South Africa, 1990–2002." *African Sociological Review* 5, no. 2 (2001): 15–35. http://www.ukzn.ac.za/ccs/default.asp?6,20,10,1809 (accessed July 14, 2007).

Coetzee, J. M. *Age of Iron*. New York: Vintage Books, 1990.

———. *Doubling the Point: Essays and Interviews*. Edited by David Attwell. Cambridge: Harvard University Press, 1992.

———. *Dusklands*. New York: Penguin, 1983 [1974].

———. *Life & Times of Michael K*. New York: Viking, 1984 [1983].

———. *White Writing: On the Culture of Letters in South Africa*. New Haven, CT: Yale University Press, 1988.

Davis, Mike. *Planet of Slums*. New York: Verso, 2006.

Department of Environmental Affairs and Tourism, South African Government. "The Polokwane Declaration on Waste Management." The First National Waste Summit, Hosted by the Department of Environmental Affairs and Tourism, Pietersburg, September 26–28, 2001. http://www.environment.gov.za/ProjProg/WasteMgmt/Polokwane_declare.htm (accessed June 6, 2008).

Desmond, Cosmas. *The Discarded People: An Account of African Resettlement in South Africa*. Harmondsworth, UK: Penguin, 1970.

Engle, Lars. "Western Classics in the South African State of Emergency: Gordimer's *My Son's Story* and Coetzee's *Age of Iron*." In *Thresholds of Western Culture: Identity, Postcoloniality, Transnationalism*, edited by John Burt Foster Jr. and Wayne Jeffrey Froman, 114–30. New York: Continuum, 2002.

Environmental and Social Report 2004. Sandton, South Africa, 2004. Toyota South Africa. www.toyota.co.za/toyotaworld/enviro_report_2003_4.pdf (accessed June 11, 2008).

Fanon, Frantz. *The Wretched of the Earth.* Translated by Constance Farrington. New York: Grove, 1963 [1961].

Hallowes, David, and Mark Butler. "Power, Poverty, and Marginalized Environments: A Conceptual Framework." In *Environmental Justice in South Africa,* edited by David A. McDonald, 51–77. Athens: Ohio University Press, 2002.

Head, Dominic. "The (Im)Possibility of Ecocriticism." In *Writing the Environment: Ecocriticism and Literature,* edited by Richard Kerridge and Neil Sammells, 27–39. New York: Zed Books, 1998.

———. *J. M. Coetzee.* New York: Cambridge University Press, 1997.

Heyns, Michiel. "An Ethical Universal in the Postcolonial Novel—'A Certain Simple Respect'?" In *Thresholds of Western Culture: Identity, Postcoloniality, Transnationalism,* edited by John Burt Foster Jr. and Wayne Jeffrey Foreman, 103–13. New York: Continuum, 2002.

———. "Houseless Poverty in the House of Fiction: Vagrancy and Genre in Two Novels by J. M. Coetzee." *Current Writing: Text and Reception in Southern Africa* 11, no. 1 (1999): 20–35.

Hoegberg, David E. " 'Where Is Hope?': Coetzee's Rewriting of Dante in *Age of Iron.*" *English in Africa* 25, no. 16 (1998): 27–42.

Huggan, Graham. "Evolution and Entropy in J. M. Coetzee's *Age of Iron.*" In *Critical Perspectives on J. M. Coetzee,* edited by Graham Huggan and Stephen Watson. London: Macmillan, 1996.

Huggan, Graham, and Helen Tiffin. "Green Postcolonialism." *Interventions* 93, no. 1 (2007): 1–11.

———. *Postcolonial Ecocriticism: Literature, Animals, Environment.* London: Routledge, 2010.

Khan, Farieda. "The Roots of Environmental Racism and the Rise of Environmental Justice in the 1990s." In *Environmental Justice in South Africa,* edited by David A. McDonald, 15–48. Athens: Ohio University Press, 2002.

———. *Waste Picking for Survival: A Report on the Waste Pickers of Frankdale Informal Settlement.* Rondebosch, South Africa: Environmental Advisory Unit, University of Cape Town, 1996.

Kuper, Jeremy. "Final Collection." *Guardian,* August 5, 2006. http://www.guardian.co.uk/money/2006/aug/05/careers.work4/print (accessed March 4, 2008).

Lakhani, Muna. "A Zero Waste Local Economy: Myth, Mystery or Magic?" Prepared for the ICLEI Conference, Cape Town, 2006. Institute for Zero Waste in Africa–IZWA, Chintan Environmental Research and Action Group. www.chintan-india.org/others/zero_waste_local_economy_iclei_booklet.pdf (accessed June 3, 2008).

Marais, Michael [Mike]. "Places of Pigs: The Tension between Implication and Transcendence in J. M. Coetzee's *Age of Iron* and *The Master of Petersburg.*" *Journal of Commonwealth Literature* 31, no. 1 (1996): 83–95.

———. "'Who Clipped the Hollyhocks?': J. M. Coetzee's *Age of Iron* and the Politics of Representation." *English in Africa* 20, no. 2 (October 1993): 1–24.

Meadows, Donella H., Dennis L. Meadows, Jørgen Randers, and William W. Behrens III. *The Limits to Growth: A Report for the Club of Rome's Project on the Predicament of Mankind.* New York: Universe Books, 1972.

Mieder, Wolfgang, Stewart A. Kingsbury, and Kelsie B. Harder, eds. *A Dictionary of American Proverbs.* New York: Oxford University Press, 2000.

Mulama, Joyce. "Menace of the Flying Toilets." Mail & Guardian Online, October 26, 2006, http://www.mg.co.za/printformat/single/2006-10-26-menace-of-the-flying-toilets (accessed July 2, 2007).

Munnik, Victor. *Learning with Mogale City: Civil Society Participation in Developing an Integrated Waste Management Plan.* Johannesburg: Group for Environmental Monitoring, 2004.

Murray, Robin. *Zero Waste.* London: Greenpeace Environmental Trust, 2002.

Myers, Garth Andrew. *Disposable Cities: Garbage, Governance and Sustainable Development in Urban Africa.* Re-Materialising Cultural Geography Series. Aldershot, UK: Ashgate, 2005.

Nixon, Rob. "Environmentalism and Postcolonialism." In *Postcolonial Studies and Beyond,* edited by Ania Loomba, Suvir Kaul, Matti Bunzl, Antoinette Burton, and Jed Esty, 233–51. Durham, NC: Duke University Press, 2005.

Ogbuokiri, Paul, and Stan Okenwa. "Toxic Ship Berths in Lagos." AllAfrica Global Media, June 4, 2010. http://allafrica.com/stories/201006040343.html (accessed June 4, 2010).

Onibokun, A. G., and A. J. Kumuyi. "Governance and Waste Management in Africa." In *Managing the Monster: Urban Waste and Governance in Africa,* edited by Adepoju G. Onibokun, 1–9. Ottawa, Ontario: International Development Research Centre, 2000.

Platzky, Laurine, and Cherryl Walker. *The Surplus People: Forced Removals in South Africa.* Johannesburg: Ravan, 1985.

Qotole, Msokoli, Mthetho Xali, and Franco Barchiesi. *The Commercialisation of Waste Management in South Africa.* Research Series #3, 2001.

Queens University, Canada, Municipal Services Project. http://www.queensu.ca/msp/pages/Project_Publications/Series/3.htm (accessed June 3, 2008).

Samson, Melanie. *Dumping on Women: Gender and Privatisation of Waste Management.* Woodstock, South Africa: Municipal Services Project and South African Municipal Workers' Union, 2003.

Summary: Human Development Report 2006; Beyond Scarcity: Power, Poverty and the Global Water Crisis. United Nations Development Programme. http://hdr.undp.org/en/media/HDR2006_English_Summary.pdf (accessed June 29, 2008).

Swilling, Mark, and David Hutt. "Johannesburg, South Africa." In *Managing the Monster: Urban Waste and Governance in Africa,* edited by Adepoju G. Onibokun, 173–225. Ottawa, Ontario: International Development Research Centre, 2000.

TVSA Forums—Premiere: Hybrid Living (SABC3). TVSA, April 30, 2008. http://www.tvsa.co.za/forum/calendar.php?do=getinfo&e=2729&day=2008-4-30&c=1 (accessed June 14, 2008).

Vital, Anthony. "Situating Ecology in Recent South African Fiction: J. M. Coetzee's *The Lives of Animals* and Zakes Mda's *The Heart of Redness.*" *Journal of Southern African Studies* 31, no. 2 (2005): 297–313.

———. "Toward an African Ecocriticism: Postcolonialism, Ecology and *Life & Times of Michael K.*" In *Research in African Literatures* 39, no. 1 (2008): 87–106. Corrected version http://www.postcolonial-ecology.net (accessed July 20, 2008).

Vital, Anthony, and Hans-Georg Erney, eds. *Postcolonial Studies and Ecocriticism.* Special issue of *Journal of Commonwealth and Postcolonial Studies* 13, no. 2, and 14, no. 1 (2006–7).

Whittaker, Mark. "Why Uganda Hates the Plastic Bag." BBC News, June 20, 2007. http://news.bbc.co.uk/1/hi/programmes/from_our_own_correspondent/6253564.stm (accessed July 2, 2007).

Wines, Michael. "Shantytown Dwellers in South Africa Protest Sluggish Pace of Change." *New York Times,* December 25, 2005. http://www.nytimes.com/2005/12/25/international/africa/25durban.html (accessed December 27, 2005).

Wright, Laura. "Displacing the Voice: South African Feminism and J. M. Coetzee's Female Narrators." *African Studies* 67, no. 1 (2008): 11–32.

Zero Waste Alliance, a Program of the International Sustainable Development Foundation. http://www.zerowaste.org/case.htm (accessed March 3, 2008).

Chapter 9

Never a Final Solution

Nadine Gordimer
and the Environmental Unconscious

Byron Caminero-Santangelo

IN HIS seminal work *Justice, Nature & the Geography of Difference,* the geographer David Harvey draws on the gap between Raymond Williams's cultural theory and the treatment of the concepts of "place, space, and environment" in his novels in order to emphasize "the difficulty in getting this tripartite conceptual apparatus into the heart of cultural theory."[1] The challenge in working with such an "apparatus" is to negotiate among different "kinds and levels of abstraction" (42): place and space, the local and the global, social processes and environment. Harvey claims that Williams chooses the novel "as a vehicle to explore these themes" because fiction "is not subject to closure in the same way that more analytic forms of thinking are" (28). In his own book, Harvey sets for himself the challenge of theorizing "these themes" without relying on idealist categories. His comments about Williams's novels suggest that fiction, which can explore "choices and possibilities, perpetually unresolved tensions and differences, subtle shifts in structures of feeling all of which stand to alter the terms of debate" (28), might be useful for such a project.

In the spirit of Harvey's observations, this chapter explores how a concept developed by ecocritic Lawrence Buell—the environmental unconscious—can help explore Nadine Gordimer's novels *The Conservationist* and *Get a Life*, and at the same time how the novels encourage an approach to Buell's concept that will resist limiting analytic closure. More specifically in terms of the latter goal, I discuss the ways that Gordimer's fiction elucidates problems in trying to formulate a category—the environment—that can be set prior to the political.

According to Buell, environmental unconscious is the result of "habitually foreshortened environmental perception."[2] The term *environmental unconscious* refers to all those aspects of the physical environment that are suppressed in our awareness by inattention, ignorance, specialized training, conventions of language, and the like. The environmental unconscious has two aspects. Its negative aspect "refers to the impossibility of individual or collective perception coming to full consciousness at whatever level: observation, thought, articulation, and so forth" (22). Its positive attribute is to be found in the potential for breakthrough achieved in bringing to awareness and then articulating what has been suppressed. Through this second attribute, the environmental unconscious can become "an enabling ground condition" for "a startling and productive" reenvisioning that enables "a fuller environmental(ist) sense of [site] than workaday perception permits" (23). In addition, I would emphasize, the notion of environmental unconscious can encourage a productive resistance to closure in environmental representation since it suggests that such representation will always necessarily entail suppression of aspects of environment.

Yet Buell's own formulation of the environmental unconscious can serve as evidence not only that all acts of environmental perception and representation involve repression in their inevitable forms of closure but also of how difficult it is to resist such closure. Buell acknowledges that he has drawn on Frederic Jameson's concept of the political unconscious, but Buell also separates the latter concept from the "environmental unconscious" by privileging "environment" as a determinate of identity. For Jameson, individual and collective identity—and, as result, all literary texts—are mediated by ideological configurations and the political relationships that generate them. Every literary text is a rewriting of a prior ideological subtext. Contrasting himself with Jameson, Buell argues that "embeddedness in spatio-physical context is even more intractably constitutive of personal and social identity, and of the ways that texts get structured, than ideology is, and very likely as primordial as unconscious psychic activity itself" (24). As Buell's reference to the "primordial" indicates, he seeks an uncovering of the ways that environment has shaped us and our understanding of the world that is more fundamental than the constitutive function of forms of consciousness determined by social roles and political relationships. This position, I will suggest, is incompatible with Nadine Gordimer's perspective even though her writing is profoundly engaged with the environmental unconscious. For Gordimer, the political fundamentally determines identity and reality, including the very shape of what might be "beyond" its influence. This is not to say that she downplays the role of "embeddedness in spatio-physical context." Ideology and environment are involved in a mutual determination in her novels. We see this in particular in the way that she constructs the relationship between the political and the natural. There is no "nature" and no concept of "nature" that can be separated from the shaping influence of ideology; at the same time, ideology can never escape the impact of the "natural." Ultimately, for Gordimer, gaining insight from the realm of "the environmental unconscious" entails a transgression of existing distinctions between nature and politics, between the environment and ideology. For Gordimer, especially in *Get a Life,* environment is "the place" that resists rather than reinscribes analytic closure.

The Conservationist offers numerous examples of what might be characterized as the operation of the environmental unconscious. The novel is focused on the processes of ego formation (and maintenance) under apartheid and on the forms of repression that these processes entail. In particular, the novel is focused on the subjectivity of Mehring, a rich white industrialist who owns a farm just outside of Johannesburg. His hegemonic perspective is often brought into question through glimpses of what he cannot entirely suppress but also what he cannot acknowledge without a substantial shift in subjectivity. These glimpses frequently entail aspects of the "spatio-physical" environment. The novel insists that the development of subjectivity must be understood in terms of the distribution of land and possessions determined by systemic relationships, especially those entailed by apartheid capitalism. Yet the material world is also precisely where the repressed lies; that which holds in place Mehring's ego is where the psychic "waste" threatens to come to the surface.

Mehring's farm is especially important in terms of understanding how the organization of land and things maintains his identity. He conceives of the farm—his "fair and lovely place"—as a spot of idyllic, managed nature separated from the dominant economic and political system through which he has made his money. For him, the farm reflects back—and naturalizes—his sense of autonomy, belonging, and uniqueness, "a sign of having remained fully human and capable of enjoying the simple things of life that poorer men can no longer afford" and of a certain freedom, "not the freedom associated with a great plane by those who long to travel, but the freedom of being down there on the earth, out in the fresh air of this place-to-get-away-to."[3] Just as he separates the farm from the system, he believes that he too escapes control. As his mistress says to him, "Ah yes, that's the trouble—you think you are inviolate" (107).

The farm is also evidence of his knowledge of and ability to manage "nature": "Reasonable productivity prevailed; he had to keep half an eye (all he could spare) on everything, all the time, to achieve even that much, and of course he had made it his business to pick up a working knowledge of husbandry, animal and crop" (23). He envisions himself as the one best able to constructively regulate "nature" in a way that those who live on the farm cannot. He is both the bringer of order—making the farm a "going concern" (82)—and the protector of its natural beauty and purity. Through these roles, his natural right to ownership is secured. He is the land's proper steward. At the same time, through his management of nature he reveals that he rises above and is autonomous from nature as well. Ultimately his wealth and power are only signs of such unique abilities and identity.

One of the deep ironies of the novel, however, is that the farm itself can offer glimpses of a very different image of Mehring than the one he normally sees reflected back. In contrast with his perspective, the novel itself emphasizes how his identity and the function of the farm in relation to it are produced by apartheid capitalism. For example, his ownership of the farm enables him to repress precisely how his acquisition of it results from dispossession and exploitation, which have also produced the very "environmental threats" that he supposedly struggles against in his role as "conservationist." As he himself acknowledges, "To keep anything the way you like it for yourself you have to have the stomach to ignore—dead and hidden—whatever intrudes" (79). What Mehring ultimately strives to conserve is not the natural integrity of the farm but rather an organization and image of it that allow for his pleasure and reinforce his identity. The novel's antihegemonic vision is brought into focus when that which is repressed surfaces, often through Mehring's consciousness of events. Most prominent here would be the material traces of the history of dispossession and its effects, traces that he perceives as forms of environmental threat: poaching, trespassing, litter, etc. The dead and only superficially buried body is both an example of theses traces and the symbol of them. The body is of a "city slicker" (15) who has apparently been killed in some kind of township violence and then dumped on the farm. In Mehring's view, this body represents an incursion on his land, both trespassing and littering, and is a form of pollution; as such, Mehring wants it removed as quickly as possible. But the body is not removed. Instead the police come and bury it but not very well, as becomes apparent later on when it resurfaces after the storm. In the course of the novel, the body becomes an image both of Mehring's lack of mastery and of how the violence and waste produced by the system that has given Mehring his wealth are an integral, if buried, part of the farm. *The Conservationist* is focused on questioning the boundaries—both geographical and categorical—that maintain the hegemonic order and its corresponding forms of identity. In particular, the novel undoes what Anthony Vital refers to as "the sheltering distinction between inside and outside, between order and litter, between normalcy and outrage" through which colonial modernity is distanced from the waste that it creates and through which its beneficiaries are able to uphold the "value" of themselves and their places in opposition to the people and places that have been wasted (see Vital, this volume). In this sense, the body as it becomes part of the farm reflects back to Mehring precisely a part of himself that he does not want to acknowledge and that threatens his identity. The novel traces the oscillation of a coming to light of such aspects of Mehring's material environment and his efforts to suppress their disruptive potential.

Mehring's notions of nature are an important part of the epistemological ordering brought into question in the novel. In his pastoral vision, nature is a realm of eternal value and beauty, clearly bounded from social processes and from history. Nature must be carefully protected from black "others" who do not understand or appreciate it and who cannot make it productive. In the opening of the novel, Mehring comes across "Pale freckled eggs" in the possession of some of the children on the farm who are taking them "to play with" (9, 12). He conceives of their play as a transgression that endangers the guinea fowl on the farm, and, more generally, as an example of the practices of the farmworkers that threaten its beauty and value: "already the farmer has had occasion to complain about the number of dogs they are harboring (a danger to the game birds)" (12). Ultimately the children's game is an image of the threat to the earth represented by such "others": "A whole clutch of guinea fowl eggs. Eleven. Soon there will be nothing left. In the country. The continent. The oceans, the sky" (11). We see this sense of threat, as well as the role that it gives him as "conservationist," repeated numerous times, as "he never leaves so much as a cigarette butt lying about to deface the farm; it's they—up at the compound—who discard plastic bags and put tins beside tree-stumps. He's forever cleaning up after them" (43). Mehring's concern reaches a kind of climax after a fire sweeps through his farm. He conceives of this fire as an invader that has come onto his land from elsewhere and ravaged his beautiful place: "The fire's territory: the invasion marked out with its inlets, promontories and beach-heads. Taken overnight" (94). He believes his farm has been permanently damaged, made ugly and less valuable. He also believes that the fire could have only come from the carelessness and ignorance of black "others" who do not understand and cannot care for nature:

> Will the willows ever be the same again? They think if the lands are saved no damage has been done. They don't understand what the vlei is, the way the vast sponge of earth held in place by the reeds in turn holds the run-off when the rains come, the way the reeds filter, shelter . . . What about the birds? Weavers? Bishop birds? Snipe? Piebald kingfisher that he sometimes sees? The duck? The guinea fowl nest in the drier sections, as well. . . . But what else—insects, larvae, the hidden mesh in there of low forms that net life, beginning small as amoeba, as the dying, rotting, beginning again? (97)

Mehring believes that he knows the operation of ecology; through this knowledge, he is able to understand the threats to the "natural" world and to protect it. In turn, the black majority's right to ownership of the land is denied, and their expulsion to places of "waste"—townships and homelands—is validated. They are the source of an environmental threat that must be contained. This perspective is the ultimate expression of what David McDermott Hughes describes as whites' efforts to secure their right to power in Africa through the discourse and practice of conservation: "For them, nature naturalizes better than empire ever did" (see Hughes, this volume).

Gordimer has herself noted that colonial conservationist thinking was one inspiration for her writing of the novel:

> As time went by, I found how it's such a paradox really because we're all for conservation; we all have this concern about the natural environment in which we live. But in the South African context, it often becomes something unpleasant and almost evil, as it did in *The Conservationist,* because there's the question of whose land? Can you own the land with a piece of paper, a deed of sale? So the concern for the birds and the beasts and the lack of concern for the human beings become another issue.[4]

As this comment suggests, the novel's focus on Mehring's conception of conservation and the notions of nature that it entails makes *The Conservationist* useful in developing a postcolonial critique of certain kinds of environmentalist ideology. The novel primarily achieves this effect through its bringing to light of an environmental unconscious that points to all that Mehring's vision represses. Despite Mehring's belief in his own cleanliness, he is, in fact, a source of trash on the farm, trash that is picked up by the farmworkers precisely because they do not see it as such. As Jacobus cleans out the ashtrays in Mehring's farmhouse, he thinks to himself that "The butts were all smoked down to precisely the same length—like the ones the children knew they must deliver to him whenever they found them in the grass" (65). The guinea fowls are not in danger from either the children or the dogs of the farmworkers; Jacobus, the leader of these workers, even tells Mehring "that there were plenty of guinea fowl about." Mehring, however, will not listen. Later he does see the birds, although he does not acknowledge his earlier mistake: "over there, over there, are twenty-three guinea fowl" (108). Similarly, the fire does not actually represent a threat to the ecosystem of which the farm is a part; the ecosystem recovers, suggesting that what he saw as ruin and waste is actually part of the very "landscape" that he deems the most valuable: "things come to life under his eyes as the syntax of a foreign language suddenly begins to yield meaning" (133). In these last two instances, Mehring does "see" what he did not before; in a sense, his ability to decipher the "foreign language" of the natural environment does advance.

However, in more substantial ways, Mehring's vision of his farm does not really change. He may "see" the guinea fowl and the recovery of the farm's flora and fauna after the fire, but he does not consider that these "sights" point to both the farmworkers' environmental knowledge and the possibility that their practices do not represent the threat that he considers them to be. He does not reflect on what the fire and its aftereffects might suggest about the "development" of the farm: that it needs neither him nor its boundaries for its protection. He also does not reevaluate what he considers waste, destruction, and decay, as opposed to conservation and growth. Finally, he does not rethink his conception of the relationship between natural processes and social processes. He remains as he is described at one point, "inattentive to the earth" (46).

In *The Conservationist*, this inattentiveness is necessary for Mehring because "the earth" challenges the strict forms of delimitation that he imposes on it and that are necessary for the maintenance of his authority and identity. In Gordimer's novel, "the earth" is the sublime, outstripping existing order and meaning. This notion is most fully figured in the storms that come in from the Mozambique Channel. These storms are awesome not only in appearance but also in their effects. They wash away roads and disrupt telephone communication, thus preventing Mehring from getting to and communicating with the farm. They also utterly transform the land itself: "The sense of perspective was changed as out on an ocean where, by the very qualification of their designation, no landmarks are recognizable" (233). The storms initially disorient Mehring because they exceed his knowledge of nature and his ability to master it. After a small culvert that he has frequently crossed overflows and finally sweeps an Afrikaner couple away to their deaths, Mehring is overwhelmed by the force involved as well as by the thought that it could have been him held helpless and killed by the stream of water. However, by imagining a conversation with a secretary about the incident (in other words, by placing it in the context of his usual roles), he recontains the threat by suggesting that he would have known better than to have crossed the gully: "But I'd never do a thing like that" (237).

He attempts this same kind of recontainment when he gets back to the farm. Because of Mehring's forced absence, Jacobus has had to take over the running of the farm entirely. As a result, Jacobus has broken from his typical role in relation to Mehring, and this disruption has the potential to challenge Mehring's sense of his own place as rightful steward. At the same time, the storm has utterly transformed the farm's natural landscape, in part because "waste" that has been buried or hidden comes to the surface. As was the case with the incident at the culvert, this physical transformation potentially points to Mehring's lack of mastery over nature. The surfacing of the man-made "trash" also suggests the connection between the farm and those places of "waste" from which, in Mehring's view, it is separated and must be protected. In the face of the changes, Mehring again tries to reassert control upon his return by giving orders and considering the ways that he will restore the farm's proper form. He thinks both that "normal procedures must be returned to" and that he must "drain the land" so that it will not be turned to "swamp" (244–245).

However, the effects of the storm cannot be so easily contained. When he is left alone amid his flooded farm, he is overwhelmed: "for the first time since the flood, he is exposed to the place, alone: it comes to him . . . in its living presence" (245). This "living presence" both smells different ("a smell of rot") and looks different: "Something heavy has dragged itself over the whole place, flattening and swirling everything." These transformations, he is aware, have been made "by an extraordinary force that has rearranged a landscape as a petrified wake." As elements of Mehring's environmental unconscious are brought to the surface, what he becomes aware of is a natural system that shapes his farm and that he cannot understand or master. In the face of this knowledge, he "feels an urge to clean up, nevertheless, although this stuff is organic; to go round collecting, as he does bits of paper or the plastic bottles they leave lying about" (246). What he has been conserving is not "nature" or "the environment" but rather his idea of them and his relationship with them. Faced with an aspect of the environment that he has never admitted, he wants to clean it up in order to be returned to himself.

Immediately afterward, Jacobus finds the body that the storm has uncovered, and Mehring jumps in his car to escape. His strong reaction suggests that more than the body has been brought to the surface: "Recognized by the shoes and apparently what's left of a face, with the—that's enough! Why hear any more, it's not going to do anybody any good. That's enough. A hundred-and-fifty thousand of them practically on the doorstep" (249). The body points to the connection between Mehring's farm and the places of waste from which he wants to separate it, the former being shaped by and feeding off the latter. He has seen himself as transcending waste and its sources, but he and his farm are part of the system that creates them. There is no escaping this relationship: "the only thing that is final is that he's [the body's] always there" (251).

As Mehring escapes from his farm in his car, he picks up a female hitchhiker who takes him to an old mine dump with the apparent intention of seducing him. In this space, Mehring once again confronts the surfacing of aspects of his environment that he has suppressed. The divisions that maintain his sense of the world and identity collapse, and this time he cannot escape. The dump, quiet, shaded by a grove of eucalyptus trees, and separated from the highway, itself initially appears as a pastoral spot. However, Mehring quickly becomes aware that it is a place covered in trash, and this brings home the fact that it is a place built from waste. He believed that he was running from one such place (his farm), an apparent retreat that is actually a manifestation of the wasteful system. However, there is no escape; he and all his places have been shaped by that system. This inability to escape is also manifested in his interactions with the woman herself. She is able to manipulate him through his body, which is another aspect of a nature that he does not understand or control, and his awareness of being watched by a man leads Mehring to believe that he has been trapped. Faced with the sense that he is in the grip of powers (social and natural) beyond him, he has a mental breakdown.

In the final scene of the novel the reader is returned to the scene of the farm, where the workers are holding a funeral for the dead body. The scene can be read as a representation of a potentially revolutionary dispensation in which the community has moved beyond the hegemonic mapping of identity by apartheid (as revealed by their sense of solidarity with the dead man, whom they initially avoided). We do not know Mehring's final fate, but this scene suggests his irrelevance in such a dispensation. It is the text's last incursion into the environmental unconscious, this time revealing a potentiality that lies in the present. The scene (and the novel as a whole) has been criticized for reinscribing a pastoral sensibility.[5] However, as Rita Barnard points out, the acceptance of the dead man (a "city slicker") as "one of them" complicates this reading, since it "undoes the opposition between city and country."[6] The scene is not a retreat from history; rather, its revolutionary potential includes its expansive representation of an environment that defies efforts to strictly separate categories (urban/rural, nature/history, environment/politics). As Barnard argues, "While the pastoral idea of the local solution is certainly expressed in the novel, the overarching artistic and ethical purpose of the text—one in which the reader is invited to participate—is to construct a new whole, by discovering the relationship between things" (78).

In many ways, then, *The Conservationist* focuses on the environmental unconscious by bringing attention to aspects of environment unrecognized by characters and, possibly, by readers as well as by drawing attention to the forces that create this unconscious. The novel also points to the transformational potential in the coming to awareness of suppressed aspects of the environment. Finally, the novel might be useful in bringing together the notion of the environmental unconscious and the political unconscious. *The Conservationist* points to the ways that environment cannot escape from being shaped by ideology as well as the ways that environment is crucial for the operation and disruption of ideology. Yet ultimately in *The Conservationist*, environment is secondary to ideology as a determination of consciousness and secondary to political relationships as a concern. (In this sense, the novel reverses rather than collapses Buell's causal priority.) Environment only moves beyond this secondary status when it takes the form of nature as the sublime, in excess of human thought and in no need of protection.

In a more recent novel, *Get a Life*, Gordimer again draws on the notion of the environmental unconscious in order to bring attention to the limitations of a prominent form of conservation. However, this time ecology and environmentalist concerns are treated more ambivalently. If they remain constrained by their connection with bourgeois subjectivity, they also serve more socially positive (possibility even transformative) functions than they do in *The Conservationist*. In many ways, the sensibility of *Get a Life* takes its tone and perspective from the central character, Paul Bannerman, "an ecologist qualified academically at universities and institutions in the USA, England, and by experience in the forests, deserts, and savannahs of West Africa and South America."[7] Through Paul's perspective, the novel emphasizes forms of ecological connection and implications of human behavior that normally remain hidden and are brought to light by the study of ecology; his training enables ways of seeing that break with the "ordinary." "The work" that he and his colleagues do "informs their understanding of the world and their place as agents within it, from the perspective that everyone, like it or not, admit it or not, acts upon the world in some way. Spray a weedkiller on this lawn and the Hoopoe delicately thrusting the tailor's needle of its beak, after insects in the grass, imbibes poison" (83). At the same time, Paul becomes aware in the course of the novel that his profession and social position necessarily impose their own limitations of vision. In particular, he realizes that his assumption that knowledge of nature and its value can be separated from political determinations results in a kind of environmental unconscious. In this sense, in *Get a Life* Gordimer remains as relentless in her focus on the ways that the natural cannot be separated from the political as she did in *The Conservationist.*

A (frightening) disruption of Paul's "ordinary" life triggers access to aspects of his environment that have been hidden from him. The novel begins with his return to his childhood home after surgery for thyroid cancer. Because he has undergone a radiation treatment that has made him dangerous to others, he must go to live with his parents for more than two weeks in order to protect his young son. This experience overturns Paul's normal roles, practices, and relationships. An active thirty-five-year-old professional, he is usually in the "wilderness" doing his work or at home helping to run his middle-class household. His body is normally an extension of his will, a fully known part of himself that he can control. Now, relatively helpless, he is confined to his parents' house and garden. With this radical transformation of his life comes a break from usual ways of understanding "environment."

Initially the disruption in perception is most obvious in terms of Paul's body. He begins to think of it as a "territory" (14)—an aspect of environment—that determines the self in unknown (and uncontrollable) ways. With this change in his sense of his body comes a reconsideration of his identity, including his sexual identity. When he is given a massage by a male masseur, it is supposed to help him "to know that body again" (87). However, as the masseur works, Paul becomes aware of a possibility in his body that he would never expect; he becomes erect as the masseur works on him: "that other self of a man, restored to him. Under the hands of a man."

The result is a further sense of alienation from the self that he has known and assumed: "So unquestioning about himself. This question coming now. Take what he is feeling as the last alienation of that state of existence" (88). The activation of the environmental unconscious (Paul's awareness of those aspects of his body that have remained suppressed) disrupts a fundamental component of his identity—his notion of his sexuality—and results in his feeling a kind of ultimate "alienation" from himself.

However, soon afterward he experiences another kind of "alienation," this time from his professional identity. As a conservation ecologist, he has absolute confidence in his understanding of what he strives to protect—ecosystems and ecological relationships—as well as their significance: "We work on background scientific research to make protest based on absolutely undeniable facts. Try for what's unchallengeable" (115). However, as he sits in the garden thinking about the threat to the Okavango Delta in Botswana by the building of ten dams, he begins to question what he knows about nature and humans' relationship with it. His reflections begin with a conception of environmental organization that is typical for him and that challenges conventional forms of political mapping: "The Okavango is an inland delta in Botswana, the country of desert and swamp landlocked in the middle of the breadth of South West, South, and South East Africa. That's it on the maps; nature doesn't acknowledge frontiers. Neither can ecology." His professional vision enables him to see beyond the limits imposed by national "frontiers," to map the world using ecological connections, and, by implication, to think differently about political interest and action. As he thinks further about the delta, he reflects on how it challenges even his scientific knowledge: "he realised he knew too abstractly, himself limited by professionalism itself, too little of the grandeur and delicacy, cosmic and infinitesimal complexity of an ecosystem complete as this. . . . Where to begin understanding what we've only got a computerspeak label for, *ecosystem*?" (90–91). Paul becomes aware of his own environmental unconscious, of how his profession and its jargon have limited his "understanding." He then calls his colleague in order to discuss how they can use this new insight to protect the delta; if they can emphasize the "inconceivably" maintained operation of this system, in particular "the beautifully managed balance" that keeps salt levels at acceptable levels, then they can convince others of the foolishness of the dams: "We're chronically short of water and it's not understood that this—what, phenomenon, marvel, much, much more than that—this intelligence of matter, receives, contains, processes, finally distributes the stuff God knows how far, linking up with other systems. . . . And some fucking consortium's going to drain, block and kill what's been *given*" (92–93).

Despite his insight, Paul still imagines the delta within the parameters of his profession; he thinks of its ecosystem as having reached a form of perfection ("beautifully managed balance") in terms of the development of "life" that puts its value beyond question.

However, after Paul gets off the phone, his reflections take him further; he enters "areas of thought" that "question certainties" without which he cannot "go on pursuing what" he does, "being what" he is (93). Specifically, he entertains the possibility that the meaning he ascribes to the destruction of the delta may be circumscribed and may not be the "truth" he assumed: "Maybe we see the disaster and don't, can't live long enough . . . to see the survival solution. Matter with infinite innovation has found, finds, will find, to renew its principle—life: in new forms, what we think is gone *forever*. In millennia, what does it count that the white rhino becomes extinct" (93–94). This insight has profoundly disturbing implications for Paul. The "good" that he believes he is fighting for cannot be defined in absolute terms; it is not given from elsewhere: "So, what is this kind of stuff, thinking . . . Heresy, how can it come to one who when asked, And what is your line, answers, What am I, I'm a conservationist, I'm one of the new missionaries here not to save souls but to save the earth" (94). Transgressing the parameters of thought given to him by his profession (committing "heresy"), Paul has lost his certainty. If "the earth" and "life" on it are, in a sense, like ecosystems that find ways to renew themselves after individual parts are destroyed (only now, ecosystems themselves can be thought of as those parts), then conservation work is no longer about ecological protection as a necessary good, sanctioned by a higher force ("life") that is known through ecological science. Instead, it is about defining "damage" and "threat" in terms of human interests, albeit while drawing on the findings of ecology. In this sense, Paul's insight is similar to one offered by *The Conservationist*. He is not "protecting" nature, because it is in no need of protection. Instead what he seeks to save is the "nature" constructed by ecology and conservation work and, by extension, the self entailed by his role as a "conservationist."

That Paul entertains these possibilities in a garden is fitting. In a sense, his doubts point to the possibility that the nature that he knows is necessarily gardenlike—shaped by social institutions and human desire—and that his own intervention in nature is part of that shaping. What he does as a conservationist ecologist is not necessarily a result of the will of nature but is at least partly determined by professional and other social factors. His presence and practices in the wilderness already make it not itself, a place of pure nature following a trajectory outside human history. If his knowledge gained in the "wilderness" enables a movement beyond the environmental unconscious of most people, his break from his ordinary form of thought in the garden is the opportunity for his emergence from his own environmental unconsciousness, which includes a challenge to the very notion of "wilderness."

Paul responds to his heretical thoughts by trying to incorporate them into his professional frame of reference—the discourse of ecology—and, through this incorporation, to limit the threat they represent: "Whatever 'forever' means, irrevocably lost, or surviving eternally, himself in this garden is part of the complexity, the necessity. As a spider web is the most fragile example of organisation, and the delta is the grandest. . . . all the waterways and shifting sand islands of contradictions: a condition of living" (94). One of the "contradictions" in *his* new "condition of living" includes the effort to come to terms with his doubt precisely through the vocabulary of his profession; this incorporation might transform those terms but also becomes a way to manage the doubt itself: "Always find the self calling on the terminology of the wilderness, so unjudgmental, to bring to circumstances the balm of calm acceptance. The inevitable grace, zest, in being a microcosm of the macrocosm's marvel. Doubt is part of it; the salt content" (94–95). Paul has been amazed at the way that the Okavango manages salt, both a part of the delta and that which could destroy it if in excess. Through the rest of the narrative, the reader observes both the continued seeping of doubt into Paul's reflections on his "life's work" and his efforts to manage this doubt.

Despite their unsettling implications, Paul's doubts do lead him to consider new professional strategies, even if these new strategies lead to further conceptual complications. The episode "in the garden" suggests that Paul cannot just rely on science—on "facts"—but must turn to rhetoric, which focuses on making appeals through techniques based on the unstable, treacherous world of human interests, emotions, and values. After that episode, he thinks more often and more effectively about how to "sell" conservation using such techniques. For example, at one point in developing a plan to oppose the mining project, he tells his fellow ecologists, "Co-ordinate all the organisations and groups for action, jack up overseas support." This "jack up" of support even includes "pop stars who'll compose songs for us." At the same time, he remains aware that as the project has "desperately become like any other publicity campaign," it risks being compromised (146). To rely on the rhetoric of the ad agency potentially undermines the apparent solidity of the scientific grounding that he and his colleagues use for arguments against environmentally damaging development. Furthermore, the use of advertising strategies potentially legitimates the ad industry, which thrives on and is in league with the forms of development they oppose.

However, if Paul is aware of the contradictions involved in his efforts to use an approach shaped by the marketing industry, he remains blind through most of the narrative to the complications resulting from his lifelong privileged position as a middle-class white man in South Africa. In his final reflections on his professional activity, Paul finally begins to confront this issue. He returns to his "heresy" by focusing on the significance of human interest for the definition of an "environmental problem" and therefore for the meaning of his work. He is thinking again about the impact of the building of the dams in the Okavango. He does so in terms of a "human reality" that has to be understood in terms of the shifting grounds of "need" as determined by position and perspective, "however you're seen or you see yourself, the immediate, market reality—that's what counts in what you learn from the mother of your children . . . is the real world" (182). Paul's wife is a marketing executive, working in a world in which the focus is on the knowledge, manipulation, and production of fear and desire. For her, the pragmatic "real world" is opposed to the "innocent environment" of the "wilderness" in which he works (153). If the "reality" from which he has worked has been semidivine, the "eternal" reality of nature, hers is a limited human one: "People don't live eternity; they live a finite Now" (182).

This line of thought takes him to another environmental threat that he and his colleagues are challenging: an Australian mining company's efforts to secure mining rights for the dunes in an ecologically sensitive area of the Wild Coast. The company has offered 15 percent profits to local people in order to secure this contract. Paul recognizes that this deal and the toll road that will accompany it can be perceived as bringing financial benefits, the kind of benefits he already enjoys:

> No-one can disagree with the necessity for blacks to enter the development economy at a major level, fifteen percent is a good start? . . . There's also the concomitant reality that a toll highway carrying the derived minerals and ilmenite . . . might bring a weekly wage to replace the sacrifice, God's gift of a few crop fields, unique endemism, and 22 kilometers of sand dunes which used to be fished from instead of mined. Bring hi-fi systems and cars. Yes! Easy to sneer at materialism and its Agency seductions while existence within it has the luxury of dissatisfaction, the wilderness to oppose it.
>
> Who's to decide. (183)

In Paul's ultimate expression of doubt, he acknowledges the possibility that what he has always considered the objective, natural "reality," the "wilderness," as well as its significance, an unchallengeable moral "good," could be the product of his own positioning within "materialism and its Agency seductions" that generate, among many luxuries, a possibility of escape into a "pure" nature. If this is the case, the "reality" that determines decisions ("Who's to decide") regarding issues of development and conservation cannot transcend the influence of socioeconomic forces. As an ecologist, he may be able to make informed claims about the ecological impact of human action, yet arguments about the significance of this impact and about a corresponding course of action will be shaped by the "human reality" of unequal economic and political relationships and the perspectives they entail. A person has not escaped the problem of social justice even—perhaps especially—when he or she has embraced the supposedly extrahuman world of the wilderness.

At this point, Paul pulls back from his observations and returns to the "reality" that he shares with his colleagues as they contemplate their next strategy. Interestingly, in this effort he turns to a gendered discourse of separate spheres of wilderness and garden, nature and culture: "This kind of research has no place in this room with [his fellow ecologists] with whom is shared what the self pursues as reality. She. Benni [his wife], it must be allowed, is the other reality. . . . Hers, chosen, or advised by its effectiveness in the finite. . . . This kind of subject is best left in the garden" (183). To continue his work as a conservation ecologist, he must work within the "reality" of a nature to be known objectively and from which will come the grounds for conservation, as opposed to "reality" determined by subject position, a reality limited ("the finite") and constructed. As a result, he attempts to separate the two "realities" once again and even draws on stereotypical gendered language to achieve this effect. Yet the continual return of his "doubt" suggests a lasting change in his "reality" that cannot be reversed, and this change is marked even at the apparent moment of reversal ("what the self pursues as reality").

The novel ends with Paul reaffirming that "Final license of destruction must never be admitted, granted. That's the creed. Work to be done" (187). The problem, of course, is that the forces of "destruction" mutate in their effort to get around conservation laws and policies. In response, Paul and his colleagues must adapt and accept that there will be no final success: "Monday the four-wheel drive back to the wilderness . . . according to the week's plan of research to which there is never a final solution, ever. That's the condition on which the work goes on, will go on" (169). The questioning of "finality" echoes a primary theme of the novel as a whole, the ways that new "realities" are constantly being created from the transgression of meaning-making boundaries: "Success sometimes may be defined as a disaster put on hold. Qualified" (99). In the face of such conditions, the characters must "get a life," be ready to adapt and redefine their reality. For the characters to continue to act effectively in the world, they must be ready to resist closure and certainty, to face the tensions within categories and the connections among them. These include the categories of the political and the natural. Yet, to act a person must also put a stop to the "play" of meaning and identity and must "get a life," even if there is recognition of its contingency. Paul must hold onto his "reality," which includes confidence in the principles of ecology and the goals of ecological sustainability, even if, to remain viable, that "reality" must be questioned and transformed. This position is similar to Stuart Hall's when he claims that if agency "in any specific instance, depends on the contingent and arbitrary stop—the necessary and temporary break in the infinite semiosis of meaning," we must also remain aware of the way that "meaning continues to unfold . . . beyond the arbitrary closure which makes it, at any moment, possible."[8] In *Get a Life,* such contradictions remain an inescapable part of "life."

Not surprisingly given this perspective, the novel's stance on the study of ecology is contradictory. On the one hand, *Get a Life* embraces many of the underlying principles of ecology, in particular the importance of systems of relationships for the constitution and preservation of identity and "life" as well as the need to be aware of the potential impact of the disruption of such systems. The novel represents ecosystems themselves as among the most important of these systems. On the other hand, *Get a Life* emphasizes the *necessary* restrictions placed on the study of ecology by the forms of political organization and ideology that shape it. The novel does not suggest that the knowledge given by ecology is "false" but rather that it must be understood as limited by ecology's political unconscious. In *Get a Life,* the more someone uncovers the environmental unconscious, the more he or she must confront a conundrum in which ecology's truths—its "reality"—cannot be reduced to an ideological effect but also in which those truths are never beyond the shaping influence of ideology. This unstable zone challenges efforts to establish any final separation or priority in terms of the relationships between ecology and politics, environment and ideology.

Get a Life's complex stance on ecology as well as on conservation represents a substantial departure from *The Conservationist.* This difference is particularly evident in the contrast between the two "conservationists."

Paul's ecological "knowledge" and environmentalism, unlike Mehring's, are by no means entirely undercut by Gordimer's irony, and their significance is not restricted to the ways that they legitimate unequal socioeconomic relationships and the identities they entail. Just as importantly, in contrast with the automaton-like Mehring, who cannot accommodate challenges to the self and world he knows, Paul's crisis leads him to question his "reality" and, to some degree, to internalize the insights he gains from that interrogation. At the same time, he remains limited by his position as a white bourgeois male subject. *Get a Life* emphasizes, more than does Gordimer's earlier novel, the possibility of change for the "conservationist" figure but not the achievement of enlightenment or transformation. However, in the later novel the meaning of this figure is still opened up, as is the significance of ecology and conservation. The activation of Paul's environmental unconscious is not just a means of interrogating the values of the middle-class protagonist, not just the means to generate irony, but is also the shifting ground for new possibilities.

Delving further into this aspect of *Get a Life* can be facilitated by considering changes in Gordimer's literary project between the publication of *The Conservationist* and the publication of the later novel. For much of her career, this project involved interrogating the consciousness bequeathed to whites by apartheid, which she saw as "the final avatar"[9] and "ultimate expression"[10] of colonialism in Africa. She sought to excavate the "hierarchy of perception that white institutions and living habits implant throughout daily experience in every white, from childhood" ("Living," 265). This project, she claimed, contributed to social change in South Africa: "The expression in art of *what really exists* beneath the surface is part of the transformation of a society" (*Writing,* 131). Among the "white institutions and living habits" she saw as perpetuating colonial ideology and relationships were conservation and ecology. For her, they were inextricably tied to apartheid (particularly in terms of issues of land ownership), and the kind of false consciousness they generated needed to be undermined in order for "real" social relations and their implications ("what really exists") to become apparent.

The years following the end of apartheid have transformed Gordimer's project. She continues to explore the forms of consciousness instilled by apartheid and its legacies; however, she necessarily examines them in terms of a different political landscape. If vast inequalities based on race continue, these inequalities are no longer enforced by law, and the relationship between class and race has become substantially more unstable. Furthermore, if the state continues to protect entrenched economic interests, it cannot ignore, at least in the same way that the apartheid government did, the interests of the impoverished majorities. Finally, as South Africa has become like other liberal democratic nations, the ways that socioeconomic identities within the nation are linked with global capitalism have been highlighted. In *Get a Life*, Gordimer is more interested than she was in her apartheid writing in class identities that span the globe and in the power of global capital to shape the nation. In her depictions of foreign companies working with local entities to wrest control over development, Gordimer seems to be in some doubt that she saw the end of imperialism with the death of the apartheid state. In this situation, if the perspectives of privileged classes often coincide more with those of elites in other nations than with those of the majorities within their own nation, concern with threats to self-determination, economic well-being, and the right to a healthy environment posed by the operation of global capital can transcend lines drawn by class, race, ethnicity, and so forth at the global, national, and regional levels.

The shift in Gordimer's project resulting from her attempts to address changing political circumstances impacts her conceptualizing of conservation and ecology in *Get a Life*. In particular, as Anthony Vital argues, Gordimer offers "a profoundly contradictory" representation of the relationship between conservation work and "economic and institutional processes and structures."[11] On the one hand, she takes into account that "the natural environment becomes . . . a concern across social sectors" and, as a result, is "no longer viewed as the preoccupation of an elite rooted in a colonial past" (97). She gestures toward the ways that environmental work and legislation can be influenced by the needs of a greater percentage of the population and can combat development projects that exploit both people and places. On the other hand, Gordimer remains alert "to how ecology in the current world might relate in discomforting ways to economy" (98). For example, environmentalism can need to make its appeal in the terms that are offered by the very forces it opposes and that subordinate the struggle against environmental degradation to the interests of economic development. Environmentalism is also often still shaped by an environmental discourse that enables conservationists from the privileged classes to align themselves with "the environment" while ignoring and/or perpetuating the economic processes that have given them their advantages and that are often the underlying causes of the environmental threats against which they struggle.

Gordimer represents her ambivalent take on conservation in South Africa using Paul's character. While she depicts him as sympathetic and his work as valuable, he also points to the continued limitations imposed on ecology and environmentalism by bourgeois subjectivity. Vital argues that despite her ambivalence, Gordimer ultimately uses Paul to suggest that against the backdrop of global capital, his form of conservation "may indeed appear as not much more than a form of middle-class coping" (107). In this reading, Paul remains a type of privileged ecologist and conservationist unable to think outside an ideology of separate spheres (nature/society, private/public). His "suburban form of ecological thinking," Vital claims, is revealed to be "profoundly ideological, a thinking in the (unacknowledged) service of maintaining the social order that benefits him" (102). Paul's complicity with the forms of predatory capital prevents him from contributing to any lasting change. Gordimer, according to Vital, depicts Paul's form of conservation as having "no particular socially transformative vision to work for," and so despite its "attractive plausibility," it takes "on for the comfortable classes the character of a seemingly endless defensive action of uncertain outcome" (106).

While this reading of the novel's perspective on Paul and his work is powerful, it does tend to present him as static (as a type). Focusing on the theme of the environmental unconscious in *Get a Life* and on Paul's incrementally changing perspective yields a somewhat different and more optimistic reading of his character and its significance. If as a result of his commitments, the changes in his "life," and the subsequent activation of his environmental unconscious he cannot permanently suppress a growing awareness of the impossibility of separating conservation and social "interests" as well as ecology and rhetoric, then his work and even some of the institutions with which he is associated have the potential to be transformed and could become even more useful in creating a better future. *He* is part of the current circumstances from which a new order might be born, unlike Mehring whose *absence* at the end of *The Conservationist* is one means by which the text gestures toward a better future arising from the circumstances of the present.

Notes

1. David Harvey, *Justice, Nature & the Geography of Difference* (Malden, MA: Blackwell, 1996), 44.

2. Lawrence Buell, *Writing for an Endangered World* (Cambridge: Harvard University Press, 2001), 18.

3. Nadine Gordimer, *The Conservationist* (New York: Penguin, 1972), 22–23.

4. Nancy Bazin and Marilyn Seymour, *Conversations with Nadine Gordimer* (Jackson: Mississippi University Press, 1990), 286–87.

5. See Irene Gorak, "Libertine Pastoral: Nadine Gordimer's *The Conservationist*," *Novel* 24, no. 3 (1992): 241–56.

6. Rita Barnard, *Apartheid and Beyond* (New York: Oxford University Press, 2007), 91. See also Kathrin Wagner, *Rereading Nadine Gordimer* (Bloomington: Indiana University Press, 1994).

7. Nadine Gordimer, *Get a Life* (New York: Penguin, 2005), 6.

8. Stuart Hall, "Cultural Identity and Diaspora," *Colonial Discourse and Post-Colonial Theory*, edited by Patrick Williams and Laura Chrisman (New York: Columbia University Press, 1994), 397.

9. Nadine Gordimer, *Writing and Being* (Cambridge: Harvard University Press, 1995), 133.

10. Nadine Gordimer, "Living in the Interregnum," in *The Essential Gesture*, edited by Stephen Clingman (New York: Penguin, 1988), 262.

11. Anthony Vital, "'Another Kind of Combat in the Bush': *Get a Life* and Gordimer's Critique of Ecology in a Globalized World," *English in Africa* 35, no. 2 (2008): 97.

Bibliography

Barnard, Rita. *Apartheid and Beyond.* New York: Oxford University Press, 2007.

Bazin, Nancy, and Marilyn Seymour. *Conversations with Nadine Gordimer.* Jackson: Mississippi University Press, 1990.

Buell, Lawrence. *Writing for an Endangered World.* Cambridge: Harvard University Press, 2001.

Gorak, Irene. "Libertine Pastoral: Nadine Gordimer's *The Conservationist*." *Novel* 24, no. 3 (1992): 241–56.

Gordimer, Nadine. *The Conservationist.* New York: Penguin, 1972.

———. *Get a Life.* New York: Penguin, 2005.

———. "Living in the Interregnum." In *The Essential Gesture*, edited by Stephen Clingman, 261–84. New York: Penguin, 1988.

———. *Writing and Being.* Cambridge: Harvard University Press, 1995.

Hall, Stuart. "Cultural Identity and Diaspora." in *Colonial Discourse and Post-Colonial Theory*, edited by Patrick Williams and Laura Chrisman, 394–403. New York: Columbia University Press, 1994.

Harvey, David. *Justice, Nature & the Geography of Difference.* Malden, MA: Blackwell, 1996.

Jameson, Fredric. *The Political Unconscious: Narrative as a Socially Symbolic Act.* Ithaca, NY: Cornell University Press, 1981.

Vital, Anthony. "'Another Kind of Combat in the Bush': *Get a Life* and Gordimer's Critique of Ecology in a Globalized World." *English in Africa* 35, no. 2 (2008): 89–118.

Wagner, Kathrin. *Rereading Nadine Gordimer.* Bloomington: Indiana University Press, 1994.

Chapter 10

Inventing Tradition and Colonizing the Plants

Ngugi wa Thiong'o's Petals of Blood *and Zakes Mda's* The Heart of Redness

Laura Wright

> To grasp the ambivalence of hybridity, it must be distinguished from an inversion that would suggest that the originary is, really, only an "effect." Hybridity has no such perspective of depth or truth to provide: it is not a third term that resolves the tension between two cultures.
>
> —Homi K. Bhabha, *The Location of Culture*

IN HIS famous coauthored study *The Invention of Tradition,* historian Eric Hobsbawm argues that "'traditions' which appear or claim to be old are often quite recent in origin and sometimes invented." He goes on to state that "'Invented tradition' is taken to mean a set of practices, normally governed by overtly or tacitly accepted rules and of a ritual or symbolic nature, which seek to inculcate certain values and norms of behavior by repetition, which automatically implies continuity with the past."[1]

Through repetition and lack of variation, Hobsbawm asserts, ritual becomes codified as tradition in that it ultimately becomes linked with past action, even if only because the action is repeatedly performed over a short period of time. Such a claim is clearly informed by Ernest Renan's theoretical stance on the importance of forgetting in the construction of nations,[2] as is Benedict Anderson's assertion, published the same year as

Hobsbawm's coauthored study, that the nation "is an imagined political community—and imagined as both inherently limited and sovereign."[3] I invoke these three well-known views of the importance of forgetting, invention, and imagination in the creation of tradition and national identity because of their theoretical relevance with regard to the two texts that I will examine in this chapter, Kenyan author Ngugi wa Thiong'o's *Petals of Blood* (1977) and South African author Zakes Mda's *The Heart of Redness* (2000).[4] As colonial powers imagined narratives of their own national prominence, they subsequently invented a countermythology—dependent upon a discourse of indigenous primitivism, spiritual vacuousness, and intellectual inferiority—that justified their takeover of African lands and the subordination of African peoples. Both Ngugi, who situates his narrative in recently independent Kenya, and Mda, who situates his immediately after the fall of apartheid in South Africa, write novels that imagine and mythologize the impact of precolonial pasts upon the postcolonial present.

Both novels deal very explicitly with the potentially devastating effects of capitalist-driven development of the land, particularly as a result of deforestation and the replacing of native flora with European varieties of plants in Mda's novel and the replacement of the symbolic significance of a plant-based drink in Ngugi's novel. Characters in both works either seek a return to an imagined traditional identity or struggle to define and occupy a hybrid space, and the presentation of the mortality and vulnerability of plant life functions in both texts as a mirror for nationalist survival. The destruction of indigenous plants and their displacement by alien species can be read on a very literal level as indicative of the decrease in biodiversity that results from the introduction of nonnative species—a product of the slow violence that Rob Nixon characterizes elsewhere in this volume—but such an instance can also be read as a cultural metaphor. Indigenous history, a connection to an authentic past, has been erased by an often appealing but insidious European presence in Africa, and tellingly in both works there are no hybrid plants, only nonnative species that crowd out and starve the native flora.

Cultural hybridization is equally problematic, as is symbolically manifest in indigenous characters' attempts to eradicate invasive plant species or re-create precolonial symbolic meaning. As Qukezwa says about the European inkberry bush in Mda's novel, "It kills other plants. These flowers that you like so much will eventually become berries. Each berry is a prospective plant that will kill the plant of my forefathers."[5] Similarly, Ngugi's narrator claims that an early white colonist, Lord Freeze-Kilby, planted wheat as a means of changing "Ilmorog wilderness into civilised shapes and forms"[6] that would yield European trees and crops. The search for an authentic or "traditional" identity that is somehow uncorrupted by the imposition of colonial culture in both works, then, is played out in part through attempted eradication of these invasive plant species. In the case of Freeze-Kilby, for example, the Ilmorog leaders "met and reached a decision. They set fire to the whole [wheat] field" (69), an instance that foreshadows the death by fire of three prominent neocolonial businessmen later in the narrative. Similarly, in *Heart of Redness* Qukezwa is brought before the Xhosa leaders for cutting down trees that are "foreign," the lantana and the wattle that have "come from other countries . . . to suffocate our trees" (216). Such actions, however, are ineffective as attempts at establishing various prelapsarian—and imaginary—African Edens, impossible landscapes that are somehow uncompromised by their postcolonial status.

Like the characters who imagine and reinterpret tradition in his work, Ngugi, as the first Kenyan author to write a novel in English—*Weep Not, Child* (1964)—neither invented history nor set out to reinstate some "authentic" version of the past but instead invented a literary tradition that, while undoubtedly influenced by a Western model, had no precedent. Therefore, in a very real sense, Ngugi's imagined narratives became Kenya's national story.[7] Ngugi's novels invent a literary tradition based on an oral culture's colonization by Western forces and subsequent neocolonial identity as shaped by black African greed. Mda's project in *Heart of Redness* is, in many ways, similar to Ngugi's. While Mda was not the first black South African author to write a novel in English, he does note that the end of apartheid played a significant role in his transformation from a playwright to a novelist. Mda, who initially wrote in his native Xhosa, notes in an interview with Elly Williams that during apartheid,

> we needed work that would directly talk to the people—like poetry. Our poetry is not written on the page for a solitary reader, but for performance. You go out and perform it. We needed plays because plays are immediate. . . . [W]e did not have the luxury to sit down for months on end working on one piece of work, such as a novel.[8]

Mda claims that white writers—such as J. M. Coetzee, Nadine Gordimer, and Andre Brink—wrote novels during apartheid; they had the time to do so. In an interview in *Africultures,* Mda says that the end of apartheid freed his imagination: "I see stories everywhere."[9] Mda's transition from playwright to novelist is significant in terms of the way that he, like Ngugi, has shaped the trajectory of indigenous African imagination, mythology, and history. In researching the Xhosa Cattle Killing of 1856–57 in order to write *Heart of Redness,* Mda gave equal footing to "written history as it exists in books and in the archives, but also history as it exists in the *imagination* of the people."[10] By codifying this cultural imagination, Mda generates a counternarrative about the Xhosa prophet Nongqawuse, one that subverts her reviled status as the destroyer of the Xhosa nation; in Mda's novel, it is Nongqawuse's historical significance that saves Xhosa land from environmentally destructive development.

Kenyan author Ngugi wa Thiong'o's *Petals of Blood* was the last fictional work that Ngugi wrote in English before renouncing that language in favor of writing in his native Gikuyu.[11] Of his decision to write in his native tongue, Ngugi asserts that he had an epistemological break with regard to language, and he notes that his decision to write in his native language functions as a kind of compromise between the recovery of a precolonial past and a historically informed present. He says that "but, of course, once you make that choice, it does not mean you are going to *invent* your own history or new world, so to speak. You still have to write from your social perspective."[12] In *Decolonising the Mind,* Ngugi is careful to note that his decision to write in Gikuyu is not an attempt to reclaim a precolonial past but instead to create a new medium for the dissemination of Gikuyu culture in the present political moment. Furthermore, he is astutely aware that "writing in our languages . . . will not itself bring about the renaissance in African cultures if that literature does not carry the content of our people's anti-imperialist struggles to liberate their productive forces from foreign control."[13]

In terms of Ngugi's shift from English to Gikuyu, even though he was still writing in English in *Petals of Blood,* the text itself is, according to Evan Mwangi, a hybrid and "contains numerous non-English expressions that are left either untranslated or loosely hanging in the sentence structure in a way that would suggest that the narrative's ideal reader . . . is competent in both English and Gikuyu."[14] The novel is very explicit in its support of Marxism and therefore works to "carry the content of our people's anti-imperialist struggles" against colonial domination, presented in this case as three directors of the Theng'eta Breweries and Enterprises—Chui, Kimeria, and Mzigo—who are members of the neocolonial elite. Furthermore, as Bonnie Roos notes, Ngugi's Marxist and nativist ideologies are closely bound to Frantz Fanon's idealization of the "revolution of the agricultural working masses as the people of the nation."[15] Ngugi's championing of the agrarian African multitude over the emergent and urban African middle class implies that a connection to the land is an essential element in the process of decolonization, and according to Simon Gikandi, during the 1970s Ngugi's writing was strongly influenced by his belief that literary expression generated "representations of a people's consciousness . . . that emerged out of their struggles with the environment and their efforts to create social life out of nature."[16] In that Ngugi believed that colonial rule not only destroyed culture through its imposition of the English language but also "alienated [Africans] from their environment" and "colonized their minds" (266), his Marxist vision is informed by an environmental ethos dependent upon a connection between colonized individuals and the land from which they have been alienated.

In *Petals of Blood,* this ethos is manifest most explicitly through the character of Wanja, who through her organization of a women's farming collective "brings life back to the very soil of Ilmorog" (Roos, "Re-Historicizing" 156). Alternately, the neocolonial businessmen turn the land and its people's traditional connections to it into a commodity dependent on the sale of corrupted indigenous cultural practices, particularly the drinking of Theng'eta, a drink made from indigenous plants. Theng'eta (which means "the spirit"), the recipe for which is passed from Wanja's grandmother Nyakinyua to her granddaughter, turns from a libation once drunk in moderation and on specific celebratory occasions to a vice that is marketed to a disillusioned populace, the majority of whom have not benefited from Kenya's 1963 independence. The drink, liquor distilled from an indigenous grain, fermented millet, is nothing more than an alcoholic beverage that, according to Nyakinyua, "can only poison your heads and intestines" (210). Only when another indigenous plant, the flower Theng'eta, is combined with the liquor does it gain a spiritual significance that can give its drinker second sight, but only if it is drunk "with faith and purity" (210) in one's heart.

The drink in its contemporary manifestation, however, is a symbol of the ways that the three aspects of cultural colonization of which Ngugi is the most critical—Christianity, Western education, and the English language—sever the connection between the material, environmental, and spiritual aspects of Kenyan culture. The drink combines traditional environmental and cultural elements, but its consumption is shaped by the neocolonial present in which the drinkers exist. In the present moment of the novel, therefore, the contemporary and postcolonial tradition of Theng'eta drinking is invented. During the colonial period the drinking of Theng'eta was banned, according to Nyakinyua, by the colonizers for its tendency, they believed, to encourage insubordination (205); therefore, the revival of the practice of drinking Theng'eta in postindependence Kenya is as much an attempt to celebrate the harvest in Ilmorog and reclaim a precolonial tradition as it is an affront to the colonial government that once banned the drink. But the purity and faith that Nyakinyua claims are requisite for a person to experience the full spiritual benefit of Theng'eta are impossible to reclaim from the past and impossible to maintain in the Kenya depicted in Ngugi's text, a place where the country's wealth lies in the hands of a relative few and where the natural environment is being exploited. The corruption of both the Theng'eta flower and the drink that is derived from it function symbolically in *Petals of Blood* to demonstrate the ways that, in the face of capitalist-driven development, a person can become drunk on the sale of the natural world, particularly its indigenous plants, despite the dangers—both to the self and to the environment—of such intoxication.

Much criticism of Ngugi's novel has focused on the unwieldy nature of the narrative, the didacticism of its Marxist message, and the implausibility of its characters. For example, James A. Ogude criticizes Ngugi's "romantic portrayal of working class leaders" and claims that Karega's "coming to consciousness, his apparent transformation from a black nationalist to a trade union leader embracing a socialist vision, remains unconvincing."[17] Ogude excuses these and other flaws, however, by citing Ngugi's extremely difficult and impractical task of attempting to write a coherent call for change "in the face of fragmentation, displacement and a basic absence of models to inspire his writing" (6). Glenn Hooper similarly defends Ngugi's ambitious project: "sometimes regarded as . . . too structurally complex, *Petals of Blood* is nevertheless a milestone in African and postcolonial literature, a text noted for its excoriating attacks on neocolonial corruption, as well as for its ambitions for a dignified, proletarian-led future."[18] What critics such as Ogude and Hooper tacitly recognize, I believe, is Ngugi's attempt to construct a coherent narrative from historical fragmentation in order to write a more positive (read: Marxist) vision of postindependence Kenya into being. It seems worth noting that in terms of Ngugi's use of Theng'eta as a symbol for the corruption of tradition in the face of capitalist modernity, he, as the first Kenyan author to write a novel in English, is uniquely situated to demonstrate the malleability of national narrative; the cultural meaning attached to the plant-based drink Theng'eta changes over the course of the novel because Ngugi, as creator of a Kenyan nationalist and Marxist mythology, changes it. The mythology at the center of the novel, then, is not the mythology of precolonial Kenya but is the mythology of the present as it attempts to come to terms with a profound sense of cultural loss. The changed status of the drink in this context demonstrates how the bastardization of the drinking of Theng'eta resonates with the corruption of other Gikuyu traditional practices, particularly the oath and the drinking of tea, which is presented earlier in *Petals of Blood* and which the drinking of Theng'eta explicitly mirrors. Both practices were reconceptualized in the context of the Mau Mau emergency of the 1950s, an event that immediately precedes the narrative action, as anticolonial symbols.

Brendan Nicholls's study is one of the only critical examinations of Ngugi's novel that explores environmental aspects of the work, despite what I feel to be an overt indictment of the environmental damage produced by colonialism and neocolonialism in *Petals of Blood*, a scenario that is presented by Ngugi in almost as didactic a manner as his Marxist agenda. In addition to its focus on the ways that Ngugi genders the landscape and the nation, Nicholls's essay also examines the role of the environment—particularly the forest—in terms of the protection that it offered to Mau Mau resistance during the emergency period of 1952–60. As Nixon claims elsewhere in this volume, for the Mau Mau the forest became "a place of cultural regeneration and political refusal." Nicholls notes as well that the Mau Mau relationship with the land and its animals was "strategically canny": "Amongst the fighters, animals were a tactical resource. Elephant tracks guided them across rough terrain and showed the most direct route to water. By listening attentively to forest sounds, such as bird calls or the erratic movement of frightened animals, the insurgents produced a sympathetic sensory landscape on which the dangers of attack or discovery were signaled long before they became imminent."[19]

Nicholls also notes the network of historical taboos associated among the Gikuyu with the needless slaughter of animals or of cutting down trees; such actions could result in dire military, meteorological, or cosmic circumstances (185). It is clear from these readings that the significance of "nature," like the meaning of the oath and the drinking of tea, changes in terms of its symbolic value over the course of Kenya's colonial history. In Ngugi's novel, both nation and nature are treated as feminine and are symbolically corrupted through a pervasive capitalist modernity that calls for the development and reshaping of the landscape, a modernity that Ngugi presents as synonymous with the prostitution that Wanja is forced to undertake as a last resort after her right to brew Theng'eta is taken from her.

Before looking explicitly at the ways that Ngugi depicts environmental devastation as the result of such symbolic prostitution, I want to take a brief and more historical look at Kenya's environmental ethos, particularly as it has evolved since the 1960s depicted in Ngugi's novel and particularly in terms of the role that real women have played in the environmental movement in Kenya. According to J. R. McNeill, there are two phases of environmental policies that mark the late twentieth century, the first of which began in the mid-1960s in rich countries such as the United States, with international cooperation remaining weak.[20] Despite the fact that the first international conference on the environment held in Stockholm in 1972 led to the United Nations Environment Program headquartered in Nairobi, it was not until the second phase, which began around 1980, that grassroots efforts began to shape the political climate of poorer countries such as Kenya (350).

Perhaps the best-known example of such action is the tree-planting effort of Kenya's Green Belt Movement organized by the National Council of Women of Kenya in 1977, the same year that *Petals of Blood* was published. Nixon's chapter (see this volume) deals with the Green Belt Movement in detail, so I will only touch on that movement here. According to McNeill, "From 1981 to 1987 the movement was led by a woman, a former professor of veterinary anatomy, Wangari Maathai. . . . . The Green Belt movement proved strong enough to make an impact on the land and provoke a backlash: it had planted 20 million trees in Kenya by 1993, but government spokesmen vilified Maathai" (351–52). According to Juan Martinez-Alier in his study *The Environmentalism of the Poor*, "in 1977, Maathai abandoned her university position to motivate other and less privileged women to protect and improve their environment."[21] Despite the fact that Maathai won the Nobel Peace Prize in 2004, according to Gloria Waggoner such campaigns as the Green Belt Movement and other movements that reached their peak in the late 1970s and early 1980s "were viewed by some as a narrow approach, avoiding a look at the impact of aggressive development activities."[22]

The development during this period of numerous organizations that formed their own individualized policies has resulted in a fractured environmental protection program with no central mission and infrastructure. Hence, the state of the Kenyan environment, particularly with regard to water and air pollution, has declined over the past several decades. According to Waggoner, in addition to a lack of safe drinking water and garbage collection, "between .5–1.5% of the Gross National Product (GNP) is lost annually due to soil erosion and desertification," "40% of the natural forest cover has been lost to lumbering and other economic development," and "wetlands have been drained for agricultural purposes" (78). Furthermore, according to Timothy Armstrong, "protected areas like the national parks and preserves contain less than 30% of Kenya's wildlife, and wildlife populations outside the protected areas are declining rapidly." This decline is directly attributable to the loss of habitat as a result of forest conversion for agricultural use.[23]

The landscape that is depicted in Ngugi's novel reflects many of these environmental problems, and Wanja, as a literary precursor to women such as Maathai, is positioned at the forefront of a perceived movement aimed at changing Kenya's environmental reality. The land is drought-stricken and overfarmed, the forests are in serious decline, and "polluted air had come to dominate" (*Petals of Blood*, 100) the lives of the residents of Ilmorog. Ngugi's text firmly places the blame for such issues on the institution of colonization, and the narrative's omniscient voice, focalized through Karega, notes early on that "the land seemed not to yield much and there was no virgin soil to escape to as in those days before colonialism" (9). Alongside the novel's presentation of Ilmorog as a place of natural beauty is the assertion that the area, throughout both history and legend, has "been threatened by the twin cruelties of unprepared-for vagaries of nature and the uncontrolled actions of men" (111). Those actions include the transformation of indigenous footpaths to railways and later roads—particularly the Trans-Africa highway—that led to the consumption of oil, the pollution of the air, and the transport of raw materials and people from one place to another. Furthermore, the Theng'eta flower, once a source of communal peace and well-being, is described as a flower with "petals of blood" (21) by a schoolboy who, as a result of his Western-style education, has no access to the plant's historical and cultural meaning; in this instance, Ngugi brilliantly indicts the role of English-language education in the loss of the natural world—both literally and culturally—for colonized Africans. The excessive environmental devastation that Ngugi chronicles throughout the novel is mirrored by the pervasive consumption of Theng'eta by characters in the novel, particularly Abdulla who begins to drink "to get drunk" and thus will not "know anything about whatever was happening around him" (312). Karega claims that Kenya's neocolonial state, where the majority of the wealth that once belonged to the colonial infrastructure now rests in the hands of a few members of the Kenyan African elite, prostitutes all Kenyans, and the beginnings of that prostitution are rendered symbolically in the corruption of Theng'eta. Karega says that "Only two nights ago we all drank Theng'eta together to celebrate a harvest and the successful ending of what was certainly a difficult year in Ilmorog. It was a good harvest and you'll agree with me that such a sense of common destiny, a collective spirit, is rare. That is why the old woman rightly called it a drink of peace. Now it has turned out to be a drink of strife" (240).

Because characters partake of the drink during a period of uncertainty, a historical moment during which the old colonial order has been mimicked by the neocolonial elite, the second sight that its ingestion inspires is troubling, producing visions of familial loss and alienation and visions of shadowy enemies encountered by Abdullah, who fought "in the forest" (228) during the Mau Mau emergency. Furthermore, over the course of the evening the characters, in the disclosures of their various visionary narratives, reveal information that will generate rifts among them.

At the end of the novel, Theng'eta's traditional significance has been so altered and its meaning so corrupted that the flower and the drink become items for both capitalistic and literal consumption: rather than inspiring unity and second sight, the drink becomes a divisive hindrance to any "traditional" vision of Kenyan unity and peace. This corruption arises as well from the capitalist-driven devastation of a natural world that once offered shelter to the Mau Mau revolutionaries and with which characters in Ngugi's novel have lost both contact and cultural understanding. As I stated earlier, when Munira first comes to Ilmorog, he takes the school children "out into the field to study nature, as he put it." While in the field, one child cries out "Look. A flower with petals of blood" (21). And when Munira first approaches what the reader later realizes is Theng'eta, his eyes fool him into seeing blood flowing from the flower. The indigenous plant is symbolically blood-soaked, and the re-creation of its traditional purpose later in the novel inspires murder. Subsequently, when the children begin to ask him difficult questions about the natural world and the place of humanity within that world, Munira becomes uneasy and swears "that he would never again take the children to the fields" (22).

The image of the bleeding flower is an image that Munira ascribes to Wanja, just as it is an image of the corruption and devastation of nature as a result of deforestation and the construction of the Trans-Africa highway, and his lack of knowledge about Theng'eta illustrates that Munira and his student—as well as the other characters—have distanced themselves and been distanced from, particularly by Western education, the natural world as it had been implicated in their precolonial traditions. Furthermore, it is an image of the corruption and commodification of reimagined or invented tradition of Theng'eta drinking in the neocolonial moment. Theng'eta, just like the land that is now covered with asphalt and railways, can never be, either literally or symbolically, what it was before colonial forces intruded on the landscape. Ngugi's narrative, having found no way to return to a precolonial past, turns instead to a vision of how to imagine Kenya's future, a hybrid model of Fanonist Marxism. It is Karega's socialist vision of cultural hybridity through the overthrow of "Imperialism: capitalism: landlords. . . . The system and its gods and its angels had to be fought consciously, consistently and resolutely by all the working people! . . . Tomorrow it would be the workers and the peasants leading the struggle and seizing power to overturn the system and all its prying bloodthirsty gods and gnomic angels, bringing to an end the foreign reign of the few over the many" (344). The novel ends with Karega's hope for "tomorrow . . . tomorrow" (345), a day when Africans can create "a new earth, another world" (295), from the earth that has been taken from them and a day when neocolonial divisiveness will be a thing of the past. That vision as yet remains unrealized.

Near the end of *Petals of Blood*, Wanja offers a perspective about the appropriate treatment of the past. She says that the past is important "but only as a living lesson to the present. I mean we must not preserve our past in a museum: rather, we must study it critically, without illusions, and see what lessons we can draw from it" (323). This sentiment arises from Wanja's earlier encounter with the utamaduni village, a place where women "sing native songs and dance for white tourists" and a place "with huts built as they imagine our huts looked before the Europeans came. Our utamaduni . . . a museum . . . for them to look at" (292). Wanja's criticism of this kind of cultural preservation as imaginary performance is explicitly mirrored in South African author Zakes Mda's *Heart of Redness*, set some twenty years after the events chronicled in Ngugi's novel, when the protagonist Camagu is critical of white shop owner John Dalton's suggestion that there be a cultural village at Qolorha that will have "proper isiXhosa huts. . . . Women will wear traditional isiXhosa costumes as their forebears used to wear. They will grind millet and polish the floors with cow dung. They will draw patterns on the walls with ochre of different colors. . . . Tourists will flock to watch young maidens dance and young men engage in stick fights. . . . They will learn how the amaXhosa of the wild coast live" (247).

Camagu claims that such a display is dishonest and does not represent the way that "real people in today's South Africa" (247) live, and his comments about the village echo Hobsbawm's assertions about invented tradition: Camagu says that "I have a problem with your plans. It is an attempt . . . to reinvent culture. When you excavate a . . . precolonial identity that is lost . . . are you suggesting that they currently have no culture?" (248). In Mda's novel, characters search for a middle ground between an inauthentic presentation of an inaccessible Xhosa past that existed before the historical Xhosa Cattle Killing of 1856–57 and a future shaped by the environmentally destructive development of a vacation resort and casino in Edenic Qolorha. By having Qolorha declared a National Heritage site because of its status as the place where Nongqawuse prophesied the salvation of the Xhosa more than a century earlier, the development of the resort is averted and a proposed backpacker hostel for tourists who want to enjoy the natural beauty of Qolorha is planned instead. Nongqawuse's prophecy, in its initial incarnation, led to mass starvation and loss of Xhosa land; however, in its contemporary context, Nongqawuse's legacy ironically saves Qolorha and its residents from a modern-day white invasion that would similarly displace them. But while the ecotourist model advocated by Mda's narrative is, in many ways, preferable to the construction of the casino, it is still problematic and, as Camagu recognizes, only temporary: "the whole country is ruled by greed," he thinks, and "sooner or later . . . the gambling complex shall come into being" (277) anyway. Like Ngugi's narrative, Mda's novel also situates the reader in a position of looking toward the future, but Camagu's view of that future—perhaps because Mda wrote his narrative decades after Ngugi's and therefore has seen which way the social and environmental wind is blowing—is pessimistic, while Karega's view is hopeful.

In an interview with Elly Williams, Mda characterizes his work as "postmodern pastoralism," writing whose so-called magically realistic elements are dependent upon African oral tradition and rural environments in which "people do not have that border between what is magical and what might be referred to as objective reality" (71). In *Heart of Redness*, the allegiances of the residents of rural Qolorha remain shaped by whether their ancestors were Believers or Unbelievers in the prophecy of the historical Nongqawuse, a young Xhosa girl who in 1856 convinced her people that in order to drive white settlers from South Africa, the Xhosa must slaughter all of their cattle and destroy their crops. The longevity of the division that resulted between these two groups during the cattle killing that followed the prophecy is manifest most explicitly in Mda's contemporary characters of Zim and Bhonco, two Xhosa elders who are diametrically oppositional to one another not only in regard to their beliefs about the validity of Nongqawuse's prophecy but about everything else as well. Despite the fact that the two men "never see any issue with the same eye" (37), both are descendants of a common ancestor, Xikixa, who is beheaded in the 1830s by John Dalton, a magistrate in the British Army under Sir Harry Smith.

Mda's narrative is heavily influenced by the work of J. B. Peires, particularly his definitive work on the subject of the Xhosa Cattle Killing, *The Dead Will Arise: Nongqawuse and the Great Xhosa Cattle-Killing*

Movement of 1856–7, and the novel contains factual information that is taken almost verbatim from Peires's text,[24] particularly his presentation of Nongqawuse's prophecy.[25] According to Peires, the "new people"—ancestors of the living Xhosa—appeared before the young prophet and "told her that the dead were preparing to rise again, and wonderful new cattle too, but first the people must kill their cattle and destroy their corn."[26] Peires speculates that Nongqawuse and other prophets were able to convince so many of the Xhosa to follow this dictate because of the lung sickness epidemic, most likely introduced by European Friesland bulls in 1853, that had claimed the lives of their own animals; the Xhosa believed that their animals had been polluted and that a purification ritual was needed. Peires describes the devastation that resulted from the killing in terms of loss of the 600,000 acres of Xhosa land to the British. In terms of loss of human life, Peires claims that "the only really reliable figure that we have is that the Xhosa population of British Kaffraria dropped by two-thirds between January and December of 1857, from 105,000 to 37,500, then again by another third to read a low of 25,916 by the end of 1858" (319). While he utilizes written history about the killing, Mda's fictional retelling also complicates various readings of the event as resulting from either the delusions of a young girl or the machinations of a corrupt imperial mission to undermine the Xhosa.[27] Instead, in Mda's narrative Nongqawuse is rendered as an imaginative child, the unwitting pawn in the battle between the British—as exemplified by Sir Harry Smith's assertion that "Extermination is now the only principle that guides me. I loved these people and considered them my children. But now I say exterminate the savage beasts!" (19)—and the Xhosa, desperate to the extent that they were willing to do whatever was necessary (and believe any narrative that offered hope) to eradicate the white presence from their land.

That land, in the present moment of Mda's narrative, has been compromised by both indigenous and colonial factors. J. R. McNeill discusses one of South Africa's most significant environmental issues, soil erosion, in terms of "communal land tenure and cultural attachment to cattle." He acknowledges that an indictment of Africa's indigenous peoples for environmental damage is controversial, but he asserts that "since cattle were the preferred store of wealth and sign of status, Africans had a general incentive to maximize their herds and overgraze unless societal rules discouraged it." He is careful to point out, however, that while social regulation might have worked precolonially, it was less effective given the ways that Africans were forced onto smaller and less fertile tracts of land as the result of "the various pressures of colonialism, marketization, and long-distance labor migration" (40), all of which had begun to affect the Xhosa prior to 1856. It is worth noting that in the context of an environmental discussion about the role of cattle in African society and the dangers of overgrazing, the cattle killing of 1856–57 was an attempt, according to both Peires and Mda, to purify the earth and rid it of psychic and physical pollution.[28] However, Robert Ross claims that despite the purification goal that underscored the killing, "Nongqawuse is now seen as having inaugurated, not the renewal, but the mass suicide of the amaXhosa." Ross reads the cattle killing as a historical metaphor: "The Cattle-Killing marks the end of the beginning of South African history. For the first time, an African society . . . had been broken" (53). It is not surprising then, given its profound significance both as a literal event and as a cultural metaphor, that the Xhosa Cattle Killing finds it way into literature from both South Africa and elsewhere,[29] a phenomenon studied by Jennifer Wenzel in 2009 in *Bulletproof: Afterlives of Anticolonial Prophecy in South Africa and Beyond.* In her work, Wenzel "excavates the role of the Xhosa cattle killing and its afterlives in imagining national communities and transnational networks of affiliation."[30] Through his incorporation of the killing in his novel, Mda challenges stereotypes of rural cattle-grazing Africans as destructive of nature even as he recasts the historical tensions between Believers and Unbelievers as a contemporary and fraught discourse between peoples who, in a very real sense, need the monetary benefits that the development of a casino would offer despite their desire to preserve their natural resources and landscapes.

Through multiple doublings and redoublings—as are immediately apparent in the binary of the Xhosa Cattle Killing–era characters of Twin and Twin-Twin, whose names indicate a doubled double, twins who initially are inseparable but later turn against one another—Mda's novel fractures and compounds individual perceptions of belief and unbelief in a way that exposes the dangers of dogma, liberal or conservative, in the face of development-driven environmental destruction as it affects postapartheid South Africa. Various critics read the work as a polyphonic exercise in the destabilization of the binary oppositions that Mda's text makes manifest through its examination of the ways that the concepts of belief and unbelief, fixed and immobile as a result of their historical position within the context of the cattle killing, are emptied of any real meaning in the "new" South African Qolorha of the novel. J. U. Jacobs, for example, reads the text as performative and asserts that Mda's narrative structure functions as "the fictional equivalent of Xhosa overtone singing"[31] practiced by Qukezwa that, through its intertextual resonances (Peires's aforementioned history, the title's play on Conrad's *Heart of Darkness,* and nods to the magical realism of authors such as Gabriel Garcia Márquez), palimpsest structure (as alternating between 1856 and present-day South Africa), and diglossic treatment of Xhosa words and phrases, contribute to the narrative's multiple voices. Similarly, Siphokazi Koyana reads the "multiple ways in which there is dialogue between Qolorha (place and physical landscape) and the varied meanings or interpretations of the events that occurred there in the past and those that are occurring in the present reality of the text."[32] In "Postcolonial Ecologies and the Gaze of Animals: Reading Some Contemporary Southern African Narratives," Wendy Woodward examines the way that Mda's presentation of Xhosa environmentalism within the text cannot be read as either unilaterally precolonial in nature or as purely the product of a postcolonial ethos: Mda "does not proffer an idealized or utopian record of ecocentric communication with the environment and animals, nor do the amaXhosa constitute a unifaceted community who foreground postcolonial ecologies either in the nineteenth or twentieth centuries."[33] Such readings point to the ways that Mda's text reaches beyond the various dualities—of belief/unbelief, environmentalism/development, and Xhosa/European—that underscore the narrative's engagement with Xhosa history.

According to Siphokazi Koyana, in *Heart of Redness* "the people of Qolorha succeed in preserving their unique vegetation, the spiritual sanctuaries associated with Nongqawuse, as well as the sea in which they play and from which they feed both their bodies and their collective memories" (56). In Mda's telling of the killing, the nineteenth-century Xhosa are environmentally responsible, people who pass from one generation to the next "the art of working the soil and looking after animals" (74), and the conservationist mentality is rendered as African in origin. As the contemporary John Dalton claims, the Xhosa king Sarhili, who ruled during the killing, "was a very strong conservationist. He created Manyube, a conservation area where people were not allowed to hunt or chop trees. He wanted to preserve these things for future generations" (165). According to Wendy Woodward, "reconnecting with the land . . . and recalling precolonial knowledges is a postcolonial strategy in Mda's . . . texts" (294); this strategy is manifest in Mda's depiction of Nongqawuse as an uncertain and perhaps unwilling prophet who "in the manner of all great prophets . . . [seems] confused and disoriented most of the time" (54). The girl, who rarely speaks, is situated between competing interests: the expulsion of white settlers by the Xhosa and the acquisition of Xhosa lands by the whites, particularly Sir George Grey, the governor of Cape Colony. Instead of being the bringer of Xhosa downfall, Nongqawuse is depicted as a victim in a battle—over the land and its flora and fauna—between colonizing forces and the indigenous Xhosa. The battle that raged between the Believers and the Unbelievers over whether or not to listen to Nongqawuse and kill their cattle is played out in the present day between the Unbelievers, who are in favor of the development of the resort complex, and the Believers, who oppose the environmental destruction that will accompany such a project. In *Heart of Redness*, Camagu is drawn to the beauty and magical status of Nongqawuse's Qolorha, "a place rich in wonders." He says that "The rivers do not cease flowing, even when the rest of the country knells in drought" (7). In the context of this battle, Zim's daughter Qukezwa emerges as a kind of reverse Nongqawuse, an active environmental prophet, working to destroy invasive species and preserve the natural biota and beauty of Qolorha.

The debate that initially takes place between the Believers and Unbelievers during the killing plays out on a microcosmic scale between Bhonco and Zim; Camagu is cast as an innocent bystander, a man new to the community who aligns himself, at least at first, with John Dalton who, despite his white skin, according to Zim, "is more of an umXhosa than most of us" (147). Bhonco asserts that the "Unbelievers stand for progress" and "want to get rid of this bush which is a sign of our uncivilization. We want developers to come and build the gambling city that will bring money to this community. That will bring modernity to our lives and will rid us of redness" (92). Zim counters that the tourists who come to Qolorha "steal our lizards and our birds" and "steal our aloes and our cycads and our usundu plants and our *ikhamanga* wild banana trees" (93). After he is visited by his totem, the brown snake, and after Qukezwa takes him to Nongqawuse's valley, Camagu's environmental activism begins to take shape, and he takes a stand against the building of the gambling complex. He realizes that "a project of this magnitude cannot be built without cutting down the forest of indigenous trees, without disturbing the bird life, and without polluting the rivers, the sea, and its great lagoon" (119).

A change in the relationship between Camagu and John Dalton is first apparent when Camagu criticizes Dalton's creation of communal water taps in the village. The people often do not pay for water, and Dalton is forced to close the taps. Camagu claims that "The water project is failing because it was opposed on the people" (179). Siphokazi Koyana reads Dalton's water scheme "as a clear example of the inadequacy of development strategies that aim to improve people's lives without giving them true ownership of the process of empowerment" (54). But as the people of Qolorha seek a solution that is both respectful of the environment and monetarily beneficial, Mda's narrative via Camagu illustrates the shortcomings inherent in all options: the casino-based tourist complex is untenable because building it will destroy the environment and displace the people of Qolorha from their beaches; Dalton's cultural village will exploit the people who, Xoliswa Ximiya feels, "act like buffoons for these white tourists.... Her people are like monkeys in a zoo, observed with amusement by white foreigners with John Dalton's assistance" (96); and, most significantly, Qukezwa's attempts to eradicate invasive plants—like Nongqawuse's attempts to drive white settlers into the sea—is presented, near the end of the novel, as equally improbable. Qukezwa is brought before the inkundla for chopping down trees that "are not the trees of our forefathers" (215). Such attempts to eradicate nonnative species, while informed by a conservationist mentality, nonetheless constitute an impossible undertaking in the current moment. Qukezwa, like Nongqawuse before her, offers a highly flawed and shortsighted solution to the problem of white intrusion: Because the landscape has been altered as a result of European influence, removing all invasive plant species would not only be impossible, but such action can only operate on a metaphoric level, functioning as a symbolic displacement of a firmly entrenched capitalist developmental system. As Wendy Woodward claims, "any victory over globalizing capitalism in the form of developers at Qolorha is only tenuous and contingent" (308).

The novel ends with a chapter that similarly weaves past and present together in a seamless unity, with both Nongqawuse and Camagu seemingly existing simultaneously. Camagu appeals to John Dalton to end their feud, claiming that "there is room for both the holiday camp and the cultural village at Qolorha." As Camagu leaves the hospital, he sees the "wattle trees along the road. Qukezwa taught him that these are enemy trees," and he feels glad that he lives in a place where "those who want to preserve indigenous plants and birds have won the day." But Camagu also acknowledges that the day that has been won will pass, that sooner or later "the gambling complex shall come into being" (277). If there is any hope to be had, the polyphonic and temporally simultaneous structure of the text seems to suggest, it is in the cyclical and nonlinear nature of history wherein Nongqawuse can be read at once as the cause of a people's destruction in the past and, through her cultural cache as such a figure, their salvation in the present and as the girl whose failed prophecy in 1856 is fulfilled late in the twentieth century. Nongqawuse "really sells the holiday camp" (276), Camagu claims, and such a statement indicates an uncomfortable verisimilitude between Camagu's ecotourist business venture and Dalton's invented, capitalistic model in terms of the cultural village; the only difference is the product being marketed: culture or nature. In the case of the holiday camp, the enigma of a dead teenage girl provides the draw to a beautiful, pristine, and untouched South African landscape, while in terms of the cultural village the invocation and invention of traditional Xhosa life provides the entertainment for which tourists will pay. The final paragraph, in which Qukezwa wonders how "this Heitsi" will save his people if he is afraid of the sea, forces the reader to consider which Heitsi and which Qukezwa are being depicted. Heitsi's assertion that "This boy does not belong in the sea! This boy belongs in the man village!" (277) indicates that this Heitsi exists in the present, a hybrid child who must inhabit the space of nature and culture—and negotiate the imposition of one upon the other—in order to imagine a future informed by, but perhaps not dependent upon, the past.

Notes

The epigraph at the beginning of the chapter is from Homi K. Bhabha, *The Location of Culture* (New York: Routledge, 2004), 113.

1. Eric Hobsbawm and Terence Ranger, *The Invention of Tradition* (Cambridge: Cambridge University Press, 1983), 1.

2. In his 1882 lecture "What Is a Nation?" Renan claims that "forgetfulness, and I shall even say historical error, form an essential factor in the creation of a nation; and thus it is that the progress of historical studies may often be dangerous to the nationality" (66).

3. Benedict Anderson, *Imagined Communities: Reflections on the Origins and Spread of Nationalism* (London and New York: Verso, 1983), 6.

4. Hobsbawm's and Anderson's studies, both published in 1983, fall approximately midway between the postindependence 1960s' Kenya depicted by Ngugi and the late 1990s' South Africa of Mda's novel. Renan's 1882 lecture was delivered one year after Boer forces defeated the British in South Africa in 1881 in the First Anglo-Boer War and three years before the scramble for Africa incited the Berlin Conference of 1884–85, during which Britain staked its claim over what is present-day Kenya. See R. Mugo Gatheru, *Kenya: From Colonization to Independence, 1888–1970* (Jefferson, NC: McFarland, 2005), 7.

5. Zakes Mda, *The Heart of Redness* (New York: Picador, 2000), 90.

6. Ngugi wa Thiong'o, *Petals of Blood* (New York: Penguin, 1977), 68.

7. See Anthony Arnove, "Pierre Bourdieu, the Sociology of Intellectuals, and the Language of African Literature," *Novel: A Forum on Fiction* 26, no. 3 (1993): 278–96. In terms of invention and imagination with regard to the nation, Arnove argues that Kenya itself is a fiction, a colonial construct (278).

8. Elly Williams, "An Interview with Zakes Mda," *Missouri Review* 28, no. 2 (2005): 69.

9. Zakes Mda, Interview, *Africultures* 40 (2001), http://www.africultures.com/anglais/articles_anglais/40mda.htm.

10. Williams, "An Interview with Zakes Mda," 74, my emphasis.

11. "Gikuyu" and "Kikuyu" are both acceptable spellings. I use "Gikuyu" in my study; authors whom I quote may use "Kikuyu."

12. Angela Lamas Rodrigues, "Beyond Nativism: An Interview with Ngugi wa Thiong'o," *Research in African Literatures* 35, no. 3 (2004): 163, my emphasis.

13. Ngugi wa Thiong'o, *Decolonizing the Mind: The Politics of Language in African Literature* (Portsmouth, NH: Heinemann, 1986), 29.

14. Evan Mwangi, "The Gendered Politics of Untranslated Language and Aporia in Ngugi wa Thiong'o's *Petals of Blood*," *Research in African Literatures* 35, no. 4 (2004): 66.

15. Bonnie Roos, "Re-Historicizing the Conflicted Figure of Woman in Ngugi's *Petals of Blood*," *Research in African Literatures* 33, no. 2 (2002): 155.

16. Simon Gikandi, *Ngugi wa Thiong'o* (Cambridge: Cambridge University Press, 2001), 263.

17. James A. Ogude, "Imagining the Oppressed in Conditions of Marginality and Displacement: Ngugi's Portrayal of Heroes, Workers, and Peasants," *Wasafiri* 28 (1998): 4–5.

18. Glenn Hooper, "History, Historiography and Self in Ngugi's *Petals of Blood*," *Journal of Commonwealth Literature* 33, no. 1 (1998): 48.

19. Brendon Nicholls, "The Landscape of Insurgency: Mau Mau, Ngugi wa Thiong'o and Gender." In *Landscape and Empire, 1770–2000*, edited by Glenn Hooper (Aldershot, UK, and Burlington, VT: Ashgate, 2000), 184–85.

20. J. R. McNeill, *Something New under the Sun: An Environmental History of the Twentieth-Century World* (New York: Norton, 2000), 350.

21. Juan Martinez-Alier, *The Environmentalism of the Poor: A Study in Ecological Conflicts and Valuation* (Northampton, MA: Edward Elger, 2003), 121.

22. Gloria Waggoner, "The Environment: What Is Kenya Doing?" In *Modern Kenya: Social Issues and Perspectives*, edited by Mary Ann Watson (Lanham, MD: University Press of America, 2000), 76.

23. Timothy Armstrong, "Wildlife Conservation in Kenya." In *Modern Kenya: Social Issues and Perspectives*, edited by Mary Ann Watson (Lanham, MD: University Press of America, 2000), 90.

24. In his dedication, Mda writes that "I am grateful . . . to Jeff Peires, whose research—wonderfully recorded in *The Dead Will Arise* . . . —informed the historical events in my fiction." It is as if Mda takes Peires's suggestion in that aforementioned work: "Even if no further information can be obtained, it

must be possible to write histories of Nongqawuse from other perspectives than mine" (322).

25. In "Duplicity and Plagiarism in Zakes Mda's *Heart of Redness,*" Andrew Offenburger has claimed that Mda's novel plagiarizes Peires's work.

26. J. B. Peires, *The Dead Will Arise: Nongqawuse and the Great Xhosa Cattle-Killing Movement of 1856–7* (Johannesburg: Ravan, 1989), 311.

27. According to Robert Ross, *A Concise History of South Africa* (Cambridge: Cambridge UP, 1999), "So efficiently did [Sir George Grey, governor of Cape Colony] exploit the Cattle-Killing that many Xhosa today are convinced that Grey himself was hiding in the reeds by the Gxarha, whispering to Nongqawuse" (53).

28. I have read the literal Xhosa Cattle Killing and fictional representations of it as a form of scapegoating. For this analysis see Laura Wright, *Writing "Out of All the Camps": J. M. Coetzee's Narratives of Displacement* (New York and London: Routledge, 2006), particularly chap. 5.

29. See in particular Sindiwe Magona, *Mother to Mother* (Boston: Beacon, 1988), and African American author John Edgar Wideman, *The Cattle Killing* (New York: Houghton Mifflin, 1996).

30. Jennifer Wenzel, *Bulletproof: Afterlives of Anticolonial Prophecy in South Africa and Beyond* (Chicago: University of Chicago Press, 2009), 3.

31. J. U. Jacobs, "Zakes Mda's *The Heart of Redness:* The Novel as *Umngqokolo,*" *Kunapipi* 24, nos. 1–2 (2002): 228.

32. Siphokazi Koyana, "Qolorha and the Dialogism of Place in Zakes Mda's *The Heart of Redness,*" *Current Writing* 15, no. 1 (2003): 52.

33. Wendy Woodward, "Postcolonial Ecologies and the Gaze of Animals: Reading Some Contemporary Southern African Narratives," *JLS/TLW* 19, nos. 3–4 (2003): 308.

Bibliography

Anderson, Benedict. *Imagined Communities: Reflections on the Origins and Spread of Nationalism.* London and New York: Verso, 1983.

Armstrong, Timothy. "Wildlife Conservation in Kenya." In *Modern Kenya: Social Issues and Perspectives,* edited by Mary Ann Watson, 89–117. Lanham, MD: University Press of America, 2000.

Arnove, Anthony. "Pierre Bourdieu, the Sociology of Intellectuals, and the Language of African Literature." *Novel: A Forum on Fiction* 26, no. 3 (1993): 278–96.

Bhabha, Homi. *The Location of Culture.* London: Routledge, 1994.

Gatheru, R. Mugo. *Kenya: From Colonization to Independence, 1888–1970.* Jefferson, NC: McFarland, 2005.

Gikandi, Simon, *Ngugi wa Thiong'o.* Cambridge: Cambridge University Press, 2001.

Hall, Cheryl Jackson. "Racial and Ethnic Antagonism in Kenya." In *Modern Kenya: Social Issues and Perspectives,* edited by Mary Ann Watson, 275–301. Lanham, MD: University Press of America, 2000.

Hobsbawm, Eric, and Terence Ranger. *The Invention of Tradition.* Cambridge: Cambridge University Press, 1983.

Hooper, Glenn. "History, Historiography and Self in Ngugi's Petals of Blood." *Journal of Commonwealth Literature* 33, no. 1 (1998): 47–62.

Jacobs, J. U. "Zakes Mda's *The Heart of Redness:* The Novel as Umngqokolo." *Kunapipi* 24, nos. 1–2 (2002): 224–36.

Johnson, Joyce. "A Note on 'Theng'eta' in Ngugi wa Thiong'o's *Petals of Blood."* *World Literature Written in English* 28, no. 1 (1988): 12–15.

Kauer, Ute. "Nation and Gender: Female Identity in Contemporary South African Writing." *Current Writing* 15, no. 2 (2003): 106–17.

Koyana, Siphokazi. "Qolorha and the Dialogism of Place in Zakes Mda's *The Heart of Redness." Current Writing* 15, no. 1 (2003): 51–62.

Lloyd, David. "The Modernization of Redness." *Scrutiny2* 6, no. 2 (2001): 34–39.

MacDonald, Michael. *Why Race Matters in South Africa.* Cambridge: Harvard University Press, 2006.

Magona, Sindiwe. *Mother to Mother.* Boston: Beacon, 1998.

Martinez-Alier, Juan. *The Environmentalism of the Poor: A Study in Ecological Conflicts and Valuation.* Northampton, MA: Edward Elger, 2003.

McNeill, J. R. *Something New under the Sun: An Environmental History of the Twentieth-Century World.* New York: Norton, 2000.

Mda, Zakes. *The Heart of Redness.* New York: Picador, 2000.

———. Interview. *Africultures* 40 (2001), http://www.africultures.com/anglais/articles_anglais/40mda.htm.

Mwangi, Evan. "The Gendered Politics of Untranslated Language and Aporia in Ngugi wa Thiong'o's Petals of Blood." *Research in African Literatures* 35, no. 4 (2004): 66–74.

Nicholls, Brendon. "The Landscape of Insurgency: Mau Mau, Ngugi wa Thiong'o and Gender." In *Landscape and Empire, 1770–2000,* edited by Glenn Hooper, 177–94. Aldershot, UK, and Burlington, VT: Ashgate, 2000.

Ogude, James A. "Imagining the Oppressed in Conditions of Marginality and Displacement: Ngugi's Portrayal of Heroes, Workers, and Peasants." *Wasafiri* 28 (1998): 3–9.

———. *Ngugi's Novels and African History: Narrating the Nation.* London: Pluto, 1999.

Peires, J. B. *The Dead Will Arise: Nongqawuse and the Great Xhosa Cattle-Killing Movement of 1856–7.* Johannesburg: Ravan, 1989.

Renan, Ernest. "What Is a Nation?" In *Poetry of the Celtic Races and Other Studies,* 61–83. Port Washington, NY: Kennikat, 1970.

Rodrigues, Angela Lamas. "Beyond Nativism: An Interview with Ngugi wa Thiong'o." *Research in African Literatures* 35, no. 3 (2004): 161–67.

Roos, Bonnie. "Re-Historicizing the Conflicted Figure of Woman in Ngugi's *Petals of Blood.*" *Research in African Literatures* 33, no. 2 (2002): 154–70.

Ross, Robert. *A Concise History of South Africa.* Cambridge: Cambridge University Press, 1999.

Thiong'o, Ngugi wa. *Decolonizing the Mind: The Politics of Language in African Literature.* Portsmouth, NH: Heinemann, 1986.

———. *Petals of Blood.* New York: Penguin, 1977.

Waggoner, Gloria. "The Environment: What Is Kenya Doing?" In *Modern Kenya: Social Issues and Perspectives,* edited by Mary Ann Watson, 75–88. Lanham, MD: University Press of America, 2000.

Wenzel, Jennifer. *Bulletproof: Afterlives of Anticolonial Prophecy in South Africa and Beyond.* Chicago: University of Chicago Press, 2009.

Wideman, John Edgar. *The Cattle Killing.* New York: Houghton Mifflin, 1996.

Williams, Elly. "An Interview with Zakes Mda." *Missouri Review* 28, no. 2 (2005): 62–79.

Woodward, Wendy. "Postcolonial Ecologies and the Gaze of Animals: Reading Some Contemporary Southern African Narratives." *JLS/TLW* 19, nos. 3–4 (2003): 290–315.

Wright, Laura. *Writing "Out of All the Camps": J. M. Coetzee's Narratives of Displacement.* New York and London: Routledge, 2006.

Chapter 11

Slow Violence, Gender, and the Environmentalism of the Poor

Rob Nixon

> Ah, what an age it is
> When to speak of trees is almost a crime
> For it is a kind of silence about injustice!
>
> Bertolt Brecht, "An die Nachgeborenen" ["To Posterity"]

WE ARE accustomed to conceiving of violence in terms that are immediate, explosive, and spectacular, as erupting into instant, concentrated visibility. But as environmentalists, we need to engage the representational and strategic challenges posed by the relative invisibility of what I call slow violence, a violence that is neither spectacular nor instantaneous but instead is incremental, as its calamitous repercussions are postponed across a range of temporal scales.[1] In so doing, we can complicate conventional assumptions about violence as a highly visible act that is newsworthy because it is event focused, time bound, and targeted at a specific body or bodies. The temporal dispersion of slow violence impacts the way we perceive and respond to a variety of social afflictions—from domestic abuse to post-traumatic stress—but has especially powerful implications for environmental calamities. Hence, a major challenge facing us is how to devise arresting stories, images, and symbols adequate to the elusive violence of delayed effects. Climate change, the thawing cryosphere, toxic drift, deforestation, soil erosion, the radioactive aftermaths of wars, acidifying oceans, and a host of other slowly unfolding environmental catastrophes confront us with formidable representational obstacles that can hinder our efforts to mobilize and act decisively. Crucially, slow violence is often not just attritional but is also exponential, operating as a major threat multiplier; slow violence can fuel, long-term, proliferating conflicts wrought from desperation as the conditions for sustaining life become geometrically degraded.

If, as Edward Said notes, struggles over geography are never reducible to armed struggle but have a profound symbolic and narrative component as well and if, as Michael Watts insists, we must attend to the "violent geographies of fast capitalism," we need to supplement both of these essential injunctions with a deeper understanding of the slow violence of delayed effects that structures so many of our most consequential forgettings.[2] Violence, above all environmental violence, needs to be seen—and thought through—as a contest over time. Faulkner's dictum that "The past is never dead. It's not even past" resonates with particular force across landscapes permeated by slow violence, landscapes of temporal overspill that elude rhetorical cleanup operations with their sanitary beginnings and endings.[3] "Is the 'Post-' in 'Postcolonial' the 'Post-' in 'Postmodern'?," Kwame Anthony Appiah famously asked, and as environmentalists we might ask similarly searching questions of the "post" in postindustrial, post-Fordist, post–Cold War, and postconflict.[4] If the past of slow violence is never past, then so too the post is never fully post: industrial particulates and effluents live on in the environmental elements we inhabit and in our very bodies, which epidemiologically and ecologically are never our simple contemporaries.[5] Something similar applies to so-called postconflict societies whose leaders may annually commemorate, as marked on the calendar, the official cessation of hostilities while ongoing intergenerational slow violence (inflicted, say, by unexploded land mines or by arms dump carcinogens) may continue hostilities by other means.

Ours is an age of onrushing turbo-capitalism in which the present feels more abbreviated than it used to, at least for the world's privileged classes who live surrounded by technological time savers that often compound the sensation of being time poor. Consequently, one of the most pressing challenges of our age is how to adjust our rapidly eroding attention spans to the slow erosions of environmental justice. If under neoliberalism the gulf between enclaved rich and outcast poor has become ever more pronounced, ours is also an era of enclaved time in which speed has become for many a self-justifying, propulsive ethic that renders "uneventful" violence (to those who live remote from its attritional lethality) a weak claimant on our time. The attosecond pace of our age, with its restless technologies of infinite promise and infinite disappointment, prompts us to keep flicking and clicking distractedly in an insatiable (and often insensate) quest for quicker sensation.

Politically and emotionally, different kinds of disaster possess unequal heft. Falling bodies, burning towers, exploding heads, avalanches, and volcanoes have a visceral page-turning power that tales of slow violence, unfolding over years, decades, and even centuries, cannot match. Stories of toxic buildup, massing greenhouse gases, and accelerated species loss due to ravaged habitats may all be cataclysmic, but they are scientifically convoluted cataclysms in which casualties are postponed, often for generations. In an age when the media venerates the spectacular and when public policy is shaped primarily around perceived immediate need, how can we convert into image and narrative disasters that are slow moving and long in the making, disasters that are anonymous and star nobody, disasters that are attritional and of indifferent interest to the sensation-driven technologies of our image-world? How can we turn the long emergencies of slow violence into stories dramatic enough to rouse public sentiment and warrant political intervention in these emergencies, whose repercussions have given rise to some of the most critical challenges of our time?

To address the challenges of slow violence is to confront the dilemma that Rachel Carson faced almost half a century ago as she sought to dramatize what she called "death by indirection."[6] Carson's subjects were biomagnification and toxic drift, forms of oblique, slow-acting violence that, like climate change, pose formidable imaginative difficulties for writers and activists alike. In struggling to give shape to amorphous menace, both Carson and reviewers of *Silent Spring* resorted to a narrative vocabulary. One reviewer portrayed the book as exposing "the new, unplotted and mysterious dangers we insist upon creating all around us,"[7] while Carson herself wrote of "a shadow that is no less ominous because it is formless and obscure" (238). To confront slow violence requires, then, that we plot and give figurative shape to formless threats whose fatal repercussions are strewn across space and time. The representational challenges are acute, requiring creative ways of drawing public attention to catastrophic acts that are low in instant spectacle but high in long-term effects. To intervene representationally entails devising the iconic symbols to embody amorphous calamities and the narrative forms to infuse them with dramatic urgency.

Kenya's Green Belt Movement and Sustainable Security

Kenya's Green Belt Movement, cofounded by Wangari Maathai, serves as an animating instance of environmental activism among poor communities that have mobilized against slow violence, in this case the incremental violence of deforestation and soil erosion. At the heart of the movement's activism stand these urgent questions: What does it mean to be at risk? What does it mean to be secure? In an era when sustainability has become a buzzword, what are the preconditions for what I would call sustainable security? And how, in seeking to advance that elusive goal, can poor environmentalists—and Maathai as a writer-activist—most effectively acknowledge, represent, and counter the violence of delayed effects?

Maathai's memoir, *Unbowed*, offers us an entry point into the complex, shifting collective strategies that the Green Belt Movement devised to oppose foreshortened definitions of environmental and human security. What emerges from the Green Belt Movement's ascent is an alternative narrative of national security, one that would challenge the militaristic male version embodied and imposed by Kenya's president Daniel arap Moi during his twenty-four years of authoritarian rule from 1978 to 2002. The Green Belt Movement's rival narrative of national security sought to foreground the longer time line of slow violence, both in exposing environmental degradation and in advancing environmental recovery. At the same time, *Unbowed* provides us with an entry point into some challenging questions about the movement memoir, not least the relationship between singular autobiography and the collective history of a social movement.

The Green Belt Movement had modest beginnings. On Earth Day in 1977, Maathai and a small cohort of like-minded women planted seven trees to commemorate Kenyan women who had been environmental activists.[8] By the time Maathai was awarded the Nobel Peace Prize in 2004, the movement had created six thousand local tree nurseries and employed one hundred thousand women to plant thirty million trees, mostly in Kenya but also in a dozen other African countries.[9] The movement's achievements have been both material—providing employment while helping anchor soil, generate shade and firewood, and replenish watersheds—and symbolic, inspiring other reforestation movements across the globe. As such, the Green Belt Movement has symbolized and enacted the conviction that (as Lester Brown has stressed in another context) "a strategy for eradicating poverty will not succeed if an economy's environmental support systems are collapsing."[10]

Early on Maathai settled on the idea of tree planting as the movement's core activity, one that over time would achieve a brilliant symbolic economy, becoming an iconic act of civil disobedience as the women's efforts to help arrest soil erosion segued into a struggle against illicit deforestation perpetrated by Kenya's draconian regime. Neither soil erosion nor deforestation posed a sudden threat, but both were persistently and pervasively injurious to Kenya's long-term human and environmental prospects. The symbolic focus of mass tree plantings helped foster a broad alliance around what might be called issues of sustainable security, a set of issues crucial not just to an era of Kenyan authoritarianism but also to the very different context of post-9/11 America, where militaristic ideologies of security have disproportionately and destructively dominated public policy and debate.

The risk of ignoring the intertwined issues of slow violence and sustainable security was evident in many American responses to the March 2003 invasion of Iraq, which was widely represented as a clean strategic and moral departure from the ugly spillages of total warfare. Even many liberal commentators adhered to this view. Hendrik Hertzberg, writing in the *New Yorker,* declared that

> Whatever else can be said about the war against the Iraqi dictatorship that began on March 19th, it cannot be said that the Anglo-American invaders have pursued anything remotely resembling a policy of killing civilians deliberately. And, so far, they have gone to great tactical and technological lengths to avoid doing it inadvertently. . . . What we do not yet know is whether a different intention, backed by technologies of precision, will produce a different political result.[11]

This war, Hertzberg continued, was not the kind that "expanded the battlefield to encompass whole societies" (15). Like most American media commentators at the conflict's outset, Hertzberg bought into the idea that so-called smart bombs exhibit a morally superior intelligence.[12] What he failed to observe was that trailing behind those luminous technologies of precision streaking across the sky was the shadow of imprecision that for years, decades, and even generations will claim the lives of random civilians through the lethal legacy of depleted uranium munitions and unexploded cluster bombs. Wars that have receded into memory often continue, through their active residues, to maim and slaughter for generations. Depending on the ordnance and strategies deployed, a quick "smart" war may morph into a long-term killer, leaving behind landscapes of dragging death.

The battlefield that unobtrusively threatens to encompass whole societies is of direct pertinence to the conditions that gave rise to Kenya's Green Belt Movement. The movement emerged in response to what might be called the violence of staggered effects in relation to ecologies of scale. From the perspective of rural Kenyan women whose local livelihood has been threatened by soil erosion's slow march, what does it mean to be secure in space and time? As Maathai notes,

> During the rainy season, thousands of tons of topsoil are eroded from Kenya's countryside by rivers and washed into the ocean and lakes. Additionally, soil is lost through wind erosion in areas where the land is devoid of vegetative cover. Losing topsoil should be considered analogous to losing territory to an invading enemy. And indeed, if any country were so threatened, it would mobilize all available resources, including a heavily armed military, to protect the priceless land. Unfortunately, the loss of soil through these elements has yet to be perceived with such urgency.[13]

What is productive about Maathai's reformulation of security here is her insistence that threats to national territorial integrity—that most deep-seated rationale for war—be expanded to include threats to the nation's integrity from environmental assaults. To reframe violence in this way is to intervene in the discourse of national defense and hence in the psychology of war. Under Kenya's authoritarian regime, the prevailing response to soil erosion was a mix of denial and resignation; the damage, the loss of land, went unsourced and thus required no concerted mobilization of national resources. The violence occurred in the passive voice despite the regime's monumental resource mismanagement.

Maathai's line of reasoning here can be connected to activist writings from elsewhere in the global South, most strikingly to Vandana Shiva's advocacy for soil security as a form of environmental justice.[14] Shiva's arguments are inflected with the distinctive history in India of the Green Revolution, peasant resistance to industrial agriculture, and the battle against corporate plant patenting, but her insistence on broadening our conception of security is consistent with the stance that underlies Maathai's soil and tree politics.

Soil erosion results in part, of course, from global forms of violence, especially human-induced climate change to which rural Kenyan women contribute little and can do very little to avert. But the desert's steady seizure of once-viable fertile land also stems from local forms of slow violence—deforestation and the denuding of vegetation—and it was at those junctures that the Green Belt women found a way to exert their collective agency. As the drivers of the nation's subsistence agriculture, women inhabited most directly the fallout from an environmental violence that is low in immediate drama but high in long-term consequences.

Resource bottlenecks are difficult to dramatize and, being deficient in explosive spectacle, typically garner little media attention. Yet the bottlenecks that result from soil erosion and deforestation can fuel conflicts for decades, costing (directly and indirectly) untold lives. Certainly if we take our cues from the media, it is easy to forget that, in the words of the American agronomist Wes Jackson, "soil is as much a nonrenewable resource as oil."[15] International and intranational contests over this finite resource can destabilize whole regions. Soil security ought to be inextricable from national security policy, not least in a society such as Kenya that has lost 98 percent of its anchoring, cleansing, and cooling forest cover since the arrival of British colonialists in the late nineteenth century (Maathai, *Unbowed,* 281). Together transnational, national, and local forces—climate change, an authoritarian regime's ruthless forest destruction, and rural desperation—fueled the assault on human and environmental security that the Green Belt Movement recognized as inextricably entangled. That threat had its roots in a colonial history of developmental deforestation, most memorably evoked in Ngugi wa Thiong'o's epic novel *Petals of Blood* in which an elder remarks on how "the land was covered with forests. The trees called rain. They also cast a shadow on the land. But the forest was eaten by the railway. You remember they used to come for wood as far as here—to feed the iron thing. Aah, they only knew how to eat, how to take away everything."[16] However, despite Ngugi's forceful critique of colonial and neocolonial land politics, his novels—as Laura Wright notes elsewhere in this volume—tend to fall back on an essentialist feminizing of the soil, replete with oppositions between a precolonial virginal purity and neocolonialism as prostitution. One of the key challenges facing Maathai as a writer and an activist was how to dramatize the gendered dynamics of Kenyan land politics without submitting to the sentimental essentialism that mars Ngugi's novels. To understand the angle of her approach requires that we engage the metaphoric underpinnings of the Green Belt Movement's gender and civic politics.

The Theater of the Tree

The Green Belt Movement's achievements in engaging the violence of deforestation and soil erosion flowed from three critical strategies. First, tree planting served not only as a practical response to an attritional environmental calamity but also served to create a symbolic hub for political resistance and for media coverage of an otherwise amorphous issue. Second, the movement was able to articulate the discourse of violent land loss to a deeper narrative of territorial theft, as perpetrated first by British colonialists and later by their neocolonial legatees. Third, the Green Belt Movement made strategic use of what might be called intersectional environmentalism, broadening the movement's base and credibility by aligning itself with—and stimulating—other civil rights campaigns that were not expressly environmental, such as the campaigns for women's rights, for the release of political prisoners, and for greater political transparency.[17]

The choice of tree planting as the Green Belt Movement's defining act proved politically astute. Here was a simple and pragmatic yet powerfully figurative act that connected with many women's quotidian lives as tillers of the soil. Soil erosion and deforestation are corrosive, compound threats that damage vital watersheds, exacerbate the silting and dessication of rivers, erode topsoil, engender firewood and food shortages, and ultimately contribute to malnutrition. Maathai and her allies succeeded in using these compound threats to forge a compound alliance among authoritarianism's discounted casualties, especially marginalized women, citizens whose environmental concerns were indissociable from their concerns over food security and political accountability.

At political flashpoints during the 1980s and 1990s, these convergent concerns made the Green Belt Movement a powerful player in a broad-based civil rights coalition that gave thousands of Kenyans a revived sense of civic agency and national possibility. The movement probed and widened the fissures within the state's authoritarian structures, clamoring for answerability within what Ato Quayson, in another context, calls "the culture of impunity."[18]

The theater of the tree afforded the social movement a rich symbolic vocabulary that helped extend the movement's civic reach. Maathai recast the simple gesture of digging a hole and putting a sapling in it as a way of "planting the seeds of peace."[19] To plant trees was metaphorically to cultivate democratic change; with a slight vegetative tweak, the gesture could breathe new life into the dead metaphor of grassroots democracy. Within the campaign against one-party rule, activists could establish a ready symbolic connection between environmental erosion and the erosion of civil rights. At the heart of this symbolic nexus was a contest over definitions of growth: each tree planted by the Green Belt Movement stood as a tangible, biological image of steady sustainable growth, a dramatic counterimage to the ruling elite's kleptocratic image of "growth," a euphemism for their high-speed piratical plunder of the nation's coffers and finite natural resources. Relevant here is William Finnegan's observation, in a broader international context, that "even economic growth, which is regarded nearly universally as an overall social good, is not necessarily so. There is growth so unequal that it heightens social conflict and increases repression. There is growth so environmentally destructive that it detracts, in sum, from a community's quality of life."[20] Certainly there is something perverse about an economic order in which the unsustainable, ill-managed plunder of resources is calculated as productive growth rather than as a loss of gross national product.

Within the metaphoric groves of "growth," we have witnessed a huge spectrum of literary tree politics. Bertolt Brecht most memorably, from his Danish exile in 1939, lamented the dark times that he lived in, times of "terrible tidings:" "Ah, what an age it is / When to speak of trees is almost a crime / For it is a kind of silence about injustice!"[21] The poem that bears those words—"An die Nachgeborenen" ["To Posterity" or "To the Unborn"]—has sometimes been invoked by those who wish to distinguish the hard, clear clarion call of radical politics from the soft claims of environmentalism. Yet Brecht was clearly writing into a particular cultural moment, into an ascendant fascism, a powerful strain of blood-and-soil German romanticism implicated in Nazism's ascent. As Kenya's Green Belt Movement testifies, there are other eras when, for the sake of the unborn, we need to talk about trees with unremitting urgency; indeed, when to be silent about trees is to become complicit in an injustice to posterity.

To plant trees is to work toward cultivating change in the fullest sense of that phrase. In an era of widening social inequity and unshared growth, the replenished forest can offer an egalitarian, participatory image of growth as being sustainable over the long haul.[22] The Moi regime vilified Maathai as an enemy of growth, development, and progress, all discourses that the ruling cabal had used to mask its high-speed plunder. Saplings in hand, the Green Belt Movement returned the blighted trope of growth to its vital, biological roots.

To plant a tree is an act of intergenerational optimism, a selfless act at once practical and utopian, an investment in a communal future that the planter will not see; to plant a tree is to offer shade to unborn strangers. To act in this manner was to secede ethically from Kenya's top-down culture of ruthless short-term self-interest. (Kenyan intellectuals used to quip that under arap Moi, "L'etat c'est Moi.")[23] A social movement devoted to tree planting, in addition to regenerating embattled forests, thus also helped regenerate an endangered vision of civic time. Against the backdrop of Kenya's winner-takes-all-and-takes-it-now kleptocracy, the movement affirmed a radically subversive ethic—an ethic of selflessness—allied to an equally subversive time frame, the *longue durée* of patient growth for sustainable collective gain.

By 1998, the Moi regime had come to treat tree planting as an incendiary, seditious act of civil disobedience. That year, the showdown between the Green Belt Movement and state power came to a head over Karura Forest, covering twenty-five hundred acres. Word spread that the regime was felling swaths of the public forest, a green lung for Nairobi and a critical catchment area for four rivers (Maathai, *Unbowed*, 262). The cleared appropriated land was being sold on the cheap to cabinet ministers and other presidential cronies who planned to build luxury developments—golf courses, hotels, and gated communities—on it. Maathai and her followers, armed with nothing but oak saplings, with which they sought to begin replanting the plundered forest, were set upon by guards and goons wielding pangas, clubs, and whips. Maathai had her head bloodied by a panga; protestors were arrested and imprisoned.

The theater of the tree has accrued a host of potent valences at different points in human history: both the planting and the felling of forests have become highly charged political acts. In the England that the Puritans fled, for example, trees were markers of aristocratic privilege, and thus on numerous occasions, insurrectionists chopped or burned down those exclusionary groves. After the Restoration, notes Michael Pollan, "replanting trees was regarded as a fitting way for a gentleman to demonstrate his loyalty to the monarchy, and several million hardwoods were planted between 1660 and 1800."[24] By contrast, early American colonists typically viewed tree felling as an act of progress that could double as a way of improving the land and laying claiming to it.

Since the early 1970s, a strong but varied transnational tradition of civil disobedience has gathered force around the fate of the forest. In March 1973 a band of hill peasants in the isolated Himalayan village of Mandal devised the strategy of tree hugging to thwart loggers who had come to fell hornbeam trees in a state forest on which the peasants depended for their livelihood. This was the beginning of a succession of such protests that launched India's Chipko movement. Three years later in the Brazilian Amazon, Francisco Chico Mendes led a series of standoffs by rubber tappers and their allies who sought to arrest uncontrolled felling and burning by rancher colonists.[25] In Thailand, a Buddhist monk was jailed when he sought to safeguard trees by ordaining them, while Julia Butterfly Hill achieved celebrity visibility during her two-year tree sit to protest the clearcutting of endangered California redwoods.

What distinguished the Green Belt Movement, like the Chipko movement before it, was the way that in protesting deforestation activists went beyond what would become standard strategies of environmental civil disobedience in the global North (sit-ins, tree hugging, or chaining one-self to a tree). For the Kenyan and Indian protestors, active reforestation became the primary symbolic vehicle for their civil disobedience. Under an undemocratic dispensation, the threatened forest can be converted into a particularly dramatic theater for reviving civic agency by throwing into relief incompatible visions of public land. To Kenya's authoritarian presi-dent the forest was state-owned, and because he and his cronies treated the nation as identical to the state, he felt at license to fell national forests and sell off the nation's public land. To the activists, by contrast, the forest was not a private presidential fiefdom but instead was commonage, the indivis-ible property of the people. The regime's contemptuous looting of Karura Forest was thus read as symptomatic of a wider contempt for the rights of the poor.

The Green Belt Movement's campaign to replant Karura assumed a po-tency that reverberated beyond the fate of one particular forest; the move-ment's efforts served as a dramatic initiative to repossess for the polity not just plundered public land and resources but also plundered political agency. Outrage over the Karura assaults soon included students and other disaf-fected groups in Nairobi, and the regime was forced to suspend its attacks on both the women and the trees. In this way the theater of the tree fortified the bond between a beleaguered environment and a beleaguered polity.[26]

For those who perpetrate slow violence, their greatest ally is the pro-tracted, convoluted vapor trail of blame. If slow violence typically occurs in the passive voice—without clearly articulated agency—the attritional deforestation of Karura and other public lands offered a clearer case than, say, soil erosion of decisive accountability. The Green Belt Movement's the-ater of the tree inverted the syntax of violence by naming the agents of destruction. Through the drama of the axed tree and the planted sapling, Maathai and her allies staged a showdown between the forces of incremen-tal violence and the forces of incremental peace; in so doing they gave a symbolic and dramatic shape to public discontent over the official culture of plunder. Ultimately Maathai saw in the culture of tree planting a way of interrupting the cycle of poverty, a cycle whereby, as she put it, "poverty is both a cause and a symptom of environmental degradation."[27]

Colonialism, Mau Mau, and the Forest in National Memory

In using the theater of the forest to reanimate political debate around ideas of sustainable growth, grassroots democracy, erosion of rights, and the seeds of change, Maathai and her resource rebels also tapped into a robust national memory of popular resistance to colonialism, above all resistance to the unjust seizure of land.[28] Maathai's memoir doesn't directly engage the issue of anticolonial memory, but it an issue that is surely pertinent to the political traction that her movement attained, given the particular place of the forest in Kenya's national symbolic archive of resistance. The confrontation during Moi's neocolonial rule between the forces of deforestation and the forces of reforestation was played out against the historic backdrop of the forest as a redoubt of anticolonialism, a heroic place that, during the Mau Mau Revolt (1952–60), achieved a mythic potency among both the British colonialists and those Kenyans—primarily Kikuyu—who fought for freedom and the restitution of their land.[29]

In the dominant colonial literature about Mau Mau (political tracts, memoirs, and fiction), the forest appears as a place beyond reach of civilization, a place of atavistic savagery where "terrorists" banded together to perform degenerate rites of barbarism.[30] For those Kenyans who sought an end to their colonial subjugation, the forest represented something else entirely: it was a place of cultural regeneration and political refusal, a proving ground where resistance fighters pledged oaths of unity, above all an oath to reclaim, by force if necessary, their people's stolen land.

The forest thus became the geographical and symbolic nexus of a peasant insurrection, as a host of Kenyan writers—Meja Mwangi, Wachira, Mangua, and Ngugi wa Thiong'o among them—have all testified.[31] From an environmental perspective, *A Grain of Wheat,* Ngugi's novel of the Mau Mau Revolt, is particularly suggestive. As Byron Caminero-Santangelo observes, most of the novel's British characters work at the Githima Forestry and Agricultural Research Station, an institution whose official aims are to advance agriculture and conservation but was founded "as part of a new colonial development plan."[32] The novel unfolds in part, then, as a clash between rival cultures of nature: between nature as instrument of colonial control (under the guise of development) and nature as a sustaining animist force, an anticolonial ally of Mau Mau forest fighters pledging oaths of liberation.[33]

The gender politics of all this are complex and compelling. In the 1950s, the forest served as a bastion not just of anticolonialism but also of warrior masculinity. Thirty years later it was nonviolent women, armed only with oak saplings and a commitment to civil disobedience, who embodied the political resistance to neocolonialism. The showdown at Karura thus reprised the anticolonial history of forest resistance in a different key: now the core fighters—Maathai's "foresters without diplomas"—were female and unarmed.[34] Does this double rescripting of resistance help explain the particularly vicious backlash from Kenya's male political establishment?

Intersectional Environmentalism, Gender, and Colonial Conservation

The colonial backdrop to the achievements of the Green Belt Movement surfaces not just through the memory of Mau Mau forest fighters but also through the contrast between colonial conservation and what might be called intersectional environmentalism. Maathai was never a single-issue environmentalist: she sought from the outset to integrate and advance the causes of environmental rights, women's rights, and human rights. The Green Belt Movement emerged in the late 1970s under the auspices of the women's movement: it was through Maathai's involvement in the Kenya Association of University Women that she was first invited to join a local chapter of Environment Liaison Centre International and from there was approached by representatives from the United Nations Environmental Programme, which in turn led to ever-widening circles of international access (Maathai, *Unbowed*, 120).

Maathai's intersectional approach to environmental justice contrasted starkly with the dominant colonial traditions of conservation, which had focused on charismatic megafauna.[35] That sharply masculinist tradition—in Kenya and, more broadly, in East and southern Africa—was associated with forced removal, with colonial appropriation of land, and with an antihuman ecology. That tradition remains part of Kenya's economic legacy, a legacy associated not just with human displacement but also, in contemporary Kenya, with local exclusion from elite cultures of leisure. In ecological as in human terms, Maathai's angle of approach was not top down. Instead of focusing on the dramatic end of the biotic chain—the elephants, rhinos, lions, and leopards that have preoccupied colonial hunters, conservationists, and foreign tourists—she drew attention to a more mundane and pervasive issue: the impact of accumulative resource mismanagement on biodiversity, soil quality, food security, and the life prospects of rural women and their families.

As Fiona Mackenzie's research reveals, the grounds for such resource mismanagement were laid during the colonial era, when conservationist and agricultural discourses of "betterment" were often deployed in the service of appropriating African lands. Focusing on colonial narratives about the environment and agriculture in the Kikuyu reserves between 1920 and

1945, Mackenzie traces the effects of the colonial bureaucracy's authoritarian paternalism, of what James C. Scott calls "the imperial pretensions of agronomic science."[36] Not least among these deleterious effects was "the recasting of the gender of the Kikuyu farmer . . . through a colonial discourse of betterment that was integrally linked to the reconstruction of agricultural knowledge."[37] Thus—and this has profound consequences for the priorities of the Green Belt Movement—colonial authorities failed to acknowledge women as primary cultivators. This refusal had the effect of diminishing the deeply grounded, adaptable knowledge (both ecological and agricultural) that women had amassed.

Maathai's refusal to subordinate the interwoven questions of environmental and social justice to the priorities of either spectacular conservation or industrial agriculture has proven crucial to the long-term adaptability of the Green Belt Movement, allowing it to regenerate itself by improvising alliances with other initiatives for sustainable security and democratic transformation. Although it was the theater of tree planting that initially garnered for Maathai and her allies media attention and international support, they expanded the circles of their activism, mobilizing for campaigns that ranged from the release of Kenya's political prisoners to debt forgiveness for impoverished nations. The Green Belt Movement's intersectional strategy helped integrate issues of attritional environmental violence into a broad movement for political answerability that, in turn, helped lead to democratic elections in Kenya in 2002.

The positioning of the Green Belt Movement at the crossroads between environmental rights and women's rights makes historic sense. Women in Kenya have borne the brunt of successive waves of dispossession dating back to the late nineteenth century, when the British colonialists shifted the structures of land ownership to women's detriment. Previously land had belonged inalienably to the extended family or clan; with the introduction of colonial taxation, that same land became deeded to a male deemed to be head of the household. As taxation forced more and more Kenyans into a wage economy and as (first under colonialism and later under neocolonial structural adjustment) cash crops such as tea, coffee, and sugarcane shrank the arable land available for food production, women became disproportionately marginalized from economic power. In the resultant cash economy, men typically owned the bank accounts.[38]

Rural women suffered the perfect storm of dispossession: colonial land theft, the individualizing and masculinizing of property, and the experience of continuing to be the primary tillers of the land under increasingly inclement circumstances, including soil erosion and the stripping of the forests. As forests and watersheds became degraded, it was the women who had to walk the extra miles to fetch water and firewood, and it was the women who had to plow and plant in once rich but now denuded land where, without the anchorage of trees, topsoil was washed and blown away. In this context, the political convergence of the campaigns for environmental and women's rights in Kenya made experiential sense: women inhabited the betrayals of successive narratives of development that had brutally excluded them. The links between attritional environmental violence, poverty, and malnutrition represented a logic that they lived. Thus, when the Moi regime laid claim to Karura Forest and Uhuru Park for private development schemes, Maathai was able to mobilize women who historically had been at the raw end of plunder that benefited minute male elites, be they colonial or neocolonial in character.

It is a measure of the threat that this intersectional environmentalism posed that in 1985 the regime demanded (ultimately without success) that the women's movement and the green movement disengage from one another (Maathai, *Unbowed*, 179). What the regime foresaw was that these women tending saplings in their rural nurseries were seeding a civil rights movement that could help propel a broader campaign for an end to direct and indirect violence in the name of greater political answerability.

Yet the rise of this civil rights movement cannot be viewed solely within a national frame: local and global geopolitics also contributed in complex and often unpredictable ways. If the forces arrayed against the movement were primarily from the ruling national elite, the resources that Maathai drew on combined the meticulously local and the expansively transnational. On the one hand, the Green Belt Movement recognized that to operate in a country where sixty-two languages are spoken, it was essential to work with teams of women fluent in the local tongue, conversant with local power dynamics, and possessing local environmental knowledge. On the other hand, the movement gained indispensable traction through support from the United Nations and Scandinavian funders.

The United States played a complex role, as it would in the rise of Ken Saro-Wiwa's Movement for the Survival of the Ogoni People. One of Saro-Wiwa's primary adversaries was an American petroleum giant, Gulf Chevron, operating collaboratively with Nigerian authoritarianism, and in Kenya (a detail that Maathai omits from her memoir) the U.S. government refused to turn the screws on President Moi because the American leaders perceived him as a friendly authoritarian and a valuable ally close to the volatile Horn of Africa. Both Maathai and Saro-Wiwa traveled to the United States and drew inspiration from the civil rights and environmental campaigns that they witnessed there. That inspiration was profoundly personal but was also crucially rhetorical, granting Maathai and Saro-Wiwa a vocabulary that helped them achieve an international resonance for what might otherwise have remained obscure campaigns for environmental justice for their nation's or region's poor.

In 1960 Maathai became one of six hundred Kenyans airlifted to the United States under President John F. Kennedy's Airlift Africa program. (When Maathai published her memoir, she couldn't have foreseen how consequential that 1960 program would be: accompanying her on that airlift was a young Kenyan named Barack Obama on scholarship to the University of Hawaii.) As a beneficiary of the Kennedy airlift, Maathai got to study at a small college in Kansas; she proceeded for her graduate work to the University of Pittsburgh, and while there she was energized by listening to Martin Luther King Jr. at the height of his powers, an experience that contributed to her intersectional attitude to movement politics whereby she would envisage environmentalism as one wing of a broader civil rights campaign. The impress of America on Maathai also manifested itself in the domain of genre: while her first book, a manual on the Green Belt Movement, had a collective center, by the second book, a memoir commissioned in response to her Nobel Prize, she felt greater pressure to deliver to her American publishers a group endeavor recast as personal journey with a singular autobiographical self as its gravitational center.

Maathai was one of seven women who founded the Green Belt Movement, yet in *Unbowed* the other women never achieve any definition as characters. I observe this less as a criticism than as a way of signaling the intractable dilemmas that attend the movement memoir.[39] To underscore this point, after Nelson Mandela emerged from prison, Little Brown paid him a high six-figure advance for his autobiography. On becoming president, he predictably fell behind with his writing, so his publisher dispatched an American ghostwriter to help speed things along. The ghostwriter discovered, to Little Brown's consternation, that Mandela's autobiography had advanced with only a smattering of "I's"; his preferred default personal pronoun was "we," as in "we, the ANC." The ghostwriter was tasked with disaggregating that "we" and channeling it into an "I" story that American readers and Oprah Winfrey viewers would recognize and respond to. For Maathai, as for Mandela, the single-authored movement memoir raises profound representational dilemmas intricately entangled with transnational power imbalances in the publishing industry and also entangled with the genre expectations of projected readers, who reside mostly in the global North. Maathai's 2004 Nobel Peace Prize—and with it the publishers' investment in a celebrity memoir—intensified the pressure on the writer to recast a collective struggle in largely personal terms. To testify under such circumstances is to confirm certain genre expectations and thereby to shape the way that political movements, not least environmental justice movements, get narrated and remembered.

Environmental Agency and Ungovernable Women: Maathai and Carson

Wangari Maathai and Rachel Carson each sought, in their different cultural milieus, to shift the parameters of what is commonly perceived as violence. They devoted themselves to questioning shibboleths about development and progress, to making visible the overlooked casualties of accumulative environmental injury, and to mobilizing public sentiment—especially among women—against the self-satisfaction and profitable complicities of a male power elite. Both writer-activists questioned the orthodox, militarized vision of security as sufficient to cope with the domino effects of exponential environmental risk, not least the intergenerational risk to food security.[40] Indeed, both saw the militarization of their societies—Cold War America of the late 1950s and early 1960s and Moi's tyrannized Kenya of the 1980s and 1990s—as exacerbating the environmental degradation that threatened long-term stability (locally, nationally, and transnationally).

Retrospectively, it is easy to focus on the achievements of these two towering figures: the social movements that they helped build, the changes in legislation and public perception that they helped catalyze, Maathai's Nobel Peace Prize, and the selection of Carson's *Silent Spring* as the most influential work of nonfiction of the twentieth century. Yet it is important to acknowledge the embattled marginalization and vilification that both women had to endure at great personal cost in order to ensure that their unorthodox visions of environmental violence and its repercussions gained political traction. Their marginality was wounding but emboldening, the engine of their originality.

Carson and Maathai were multiply extrainstitutional: as female scientists (anomalies for their time and place), as scientists working outside the structures and strictures of the university, and as unmarried women. On all fronts, they had to weather *ad feminam* assaults from male establishments whose orthodoxies were threatened by their autonomy.

Although Carson had a master's degree in biology, financial pressures and the pressures of caring for dependent relatives had prevented her from pursuing a PhD. Her background was in public science writing; she had no university affiliation at a time when only 1 percent of tenured scientists in America were women.[41] But by the time she came to embark on *Silent Spring,* her best-selling books on the sea had given her some financial autonomy. Carson's institutional and economic independence freed her to set her own research agenda, to engage in unearthing, synthesizing, and promoting environmental research that had been suppressed or sidelined by the funding priorities of the major research institutions, whose agendas she recognized as compromised by the entangled special interests of agribusiness, by the chemical and arms industries, and by the headlong rush to profitable product development.

Carson's detractors questioned her professional authority, her patriotism, her ability to be unemotional, and the integrity of her scientific commitment to intergenerational genetic issues, given that she was a "spinster." "Why is a spinster with no children so concerned about genetics? She is probably a Communist," a former U.S. secretary of agriculture intoned.[42]

Hostile reviewers dismissed Carson's arguments as "hysterically over-emphatic" and as "more emotional than accurate."[43] The general counsel for Velsicol, a Chicago chemical company, accused Carson of being under the sway of "sinister influences" whose purpose was "to reduce the use of agricultural chemicals in this country and the countries of western Europe, so that our supply of food will be reduced to east-curtain parity."[44] Other commentators deduced that "Miss Rachel Carson's reference to the selfishness of insecticide manufacturers probably reflects her Communist sympathies."[45] Carson's nemesis, the chemical industry spokesman Dr. Robert White-Stevens (who gave twenty-eight speeches against *Silent Spring* in a single year), opined that "if man were to faithfully follow the teachings of Miss Carson, we would return to the Dark Ages."[46] In the ultimate vilification of Carson as embodying a model of irrational female treachery, a critic in *Aerosol Age* concluded that "Miss Carson missed her calling. She might have used her talents in telling war propaganda of the type made famous by Tokyo Rose and Axis Sally."[47]

Twenty-five years later Maathai's opponents were brandishing even more outrageous *ad feminam* threats and insinuations against an autonomous female scientist who threatened the political and environmental status quo. Maathai was not a "spinster," but she was a divorcee, a label that her opponents wielded against her relentlessly. Like Carson, Maathai was represented as overly emotional, unhinged and as an unnatural woman, uncontrollable and unattached without a husband to rein her in and keep her (and her ideas) respectable. While the chemical-agricultural establishment sought to dismiss Carson, who lacked a PhD, as unqualified to speak,

Kenya's power elite tried to discredit Maathai—the first woman in East or Central Africa to receive a doctorate in any scientific field—as suspiciously overqualified, as a woman who had to be brought down because she was overreaching.[48] When she led the protests against government plans for the private development of Uhuru Park, one parliamentarian declared that "I don't see why we should listen to a bunch of divorced women." Another politician portrayed her as a "madwoman"; a third threatened to "circumcise" her if she ever set foot in his district.[49]

As a highly educated woman scientist, an advocate of women's rights, and a proponent of environmentalism for the poor, Maathai was vulnerable on multiple fronts to charges of inauthenticity and, like Carson, of unpatriotic behavior. A Kenyan cabinet minister railed against Maathai as "an ignorant and ill-tempered puppet of foreign masters."[50] Another cabinet minister criticized her for "not being enough of an African woman," of being "a white woman in black skin" (Maathai, *Unbowed*, 110). Such critics typically adhered to a gender-specific nativism: as Maathai notes, Kenyan men freely adopted Western languages, Western dress, and the technological trappings of modernity while expecting women to be the markers and bearers of "tradition" (111).[51] President Moi (who imprisoned Maathai several times) chastised her for being "disobedient"; if she were "a proper woman in the African tradition—[she] should respect men and be quiet" (115, 196).[52]

As Kwame Anthony Appiah has observed, the charge of inauthenticity is an inherently unstable one. "Nativists may appeal to identities that are both wider and narrower than the nation: to 'tribes' and towns, below the nation-state; to Africa, above. And, I believe, we shall have the best chance of re-directing nativism's power if we challenge not the rhetoric of the tribe, the nation, or the continent, but the topology that it presupposes, the opposition it asserts."[53]

This is certainly borne out in Maathai's case: she fell foul of proliferating "uns"—unAfrican, unKenyan, unKikuyu, unpatriotic, ungovernable, unmarried, unbecoming of a woman. But through her intersectional environmentalism, she sought to circumvent the binaries of authentication. One strategy that she used to sidestep such oppositional topologies was to seek out local currents of environmental practice that were consistent with notions such as biodiversity, the commons, and ecological stewardship but not necessarily reducible to them. In this way she could also try to defuse accusations that she was a Western agent of "green imperialism."

The vehement attacks on Maathai and Carson are a measure both of institutionalized misogyny and of how much is at stake (politically, economically, and professionally) in keeping the insidious dynamics and repercussions of slow violence concealed from view. While personally vulnerable, Maathai and Carson were threatening because they stood outside powerful systems of scientific patronage, academic intimidation, and silencing kickbacks. Their cultural contexts differed widely, but their extrainstitutional positions allowed them the scientific autonomy and political integrity to speak out against attritional environmental violence and help mobilize against it.

Conclusion

While Maathai's nativist detractors sought to discredit her as an enemy of national development, she also faced, when awarded the 2004 Nobel Peace Prize, a different style of criticism from abroad. Carl I. Hagen, leader of Norway's Progress Party, typifies this line of aggressive disbelief: "It's odd," Hagen observes, "that the [Nobel] committee has completely overlooked the unrest that the world is living with daily, and given the prize to an environmental activist."[54] The implications of Hagen's position are clear: that nineteen months into the Iraq War and amid the war in Afghanistan, the wider "war on terror," and the tumult in the Middle East, Congo, Sudan, and elsewhere, to honor an environmentalist for planting trees was to trivialize conflict resolution and to turn one's back on the most urgent issues of the hour.

Maathai, however, sought to recast the question of urgency in a different time frame, one that challenged the dominant associations of two of the early twenty-first century's most explosive words: *preemptive* and *terror.* The Green Belt Movement focused not on conventional ex post facto conflict resolution but instead on conflict preemption through nonmilitary means. "Many wars in the world are actually fought over natural resources," Maathai insisted. "In managing our resources and in sustainable development . . . we plant the seeds of peace."[55] This approach has discursive, strategic, and legislative ramifications for the "global war on terror." For most of our planet's people there are more immediate terrors than a terrorist attack: creeping deserts that reduce farms to sand, the incremental assaults of climate change compounded by deforestation, not knowing where tonight's meal will come from, unsafe drinking water, or having to walk five or ten miles to collect firewood to keep one's children warm and fed. Such quotidian terrors haunt the lives of millions immiserated, abandoned, and humiliated by authoritarian rule and by a purportedly postcolonial new world order. Under such circumstances, slow violence (often coupled with direct repression) can ignite tensions, creating flashpoints of desperation and explosive rage.[56] Perhaps to Hagen and others like him, tree planting is conflict resolution lite; it lacks a dramatic, decisive, newsworthy military target. But Maathai, by insisting that resource bottlenecks impact sustainable security at local, national, and global levels and by insisting that the environmentalism of the poor is inseparable from distributive justice, has done more than forge a broad political alliance against Kenyan authoritarian rule. Through her testimony and through her movement's collective example, she has sought to reframe conflict resolution for an age when instant cinematic catastrophe has tended to overshadow violence that is calamitous in more insidious ways. This, then, is Wangari Maathai's contribution to the "war on terror": building a movement committed, in her words, to "reintroducing a sense of security among ordinary people so they do not feel so marginalized and so terrorized by the state."[57]

Notes

1. An earlier version of this chapter appeared in *Journal of Commonwealth and Postcolonial Studies* 13, no. 2 (2006–7): 14–37. I offer a fuller theoretical account of slow violence in my book *Slow Violence and the Environmentalism of the Poor* (Cambridge: Harvard University Press, 2011).

2. Edward W. Said, *Culture and Imperialism* (London: Chatto and Windus, 1992), 17, and Michael J. Watts, *Struggles over Geography: Violence, Freedom, and Development at the Millennium,* Hettner Lectures No. 3 (Heidelberg: Department of Geography, University of Heidelberg, 2000), 8.

3. Faulkner's oft-misquoted remark appears in *Requiem for a Nun* (1951; reprint, New York: Routledge, 1987), 17.

4. Anne McClintock, Aamir Mufti, and Ella Shohat, eds., *Dangerous Liaisons* (Minneapolis: University of Minnesota Press, 1997), 420–44.

5. For an astute examination of the impact of particulate residues on the time lines of environmental thinking as well as on Victorian literary genres, see Jesse Oak Taylor, "'A Sky of Our Manufacture': Literature, Modernity and the London Fog from Dickens to Conrad" (PhD diss., University of Wisconsin–Madison, 2009).

6. Rachel Carson, *Silent Spring* (1962; reprint, Boston: Houghton Mifflin, 1992), 32.

7. Eric Sevareid, "An Explosive Book," *Washington D.C. Star,* October 9, 1962, 3.

8. Anna Lappe and Frances Moore Lappe, "The Genius of Wangari Maathai," *International Herald Tribune,* October 15, 2004, 1–2.

9. Wangari Maathai, *Unbowed: A Memoir* (New York: Knopf, 2006), 175.

10. Lester R. Brown, "The Price of Salvation," *Guardian,* April 25, 2007, 1.

11. Hendrik Hertzberg, "The War in Iraq," *New Yorker,* March 27, 2003, 15.

12. The insistence that "shock and awe" was the beginning of a war unprecedented in its humanitarian precision was heard across the political spectrum. Donald Rumsfeld, most memorably, insisted that the futuristic weaponry that the United States deployed in the war exhibited "a degree of precision that no one dreamt of in a prior conflict," resulting in bombings that were morally exemplary: "The care that goes into it, the humanity that goes into it, to see that military targets are destroyed, to be sure, but that it's done in a way, and in a manner, and in a direction and with a weapon that is appropriate to that very particularized target. . . . I think that will be the case when ground truth is achieved" (United States Department of Defense, "DoD News Briefing—Secretary Rumsfeld and Gen. Myers," Friday, March 21, 2003).

13. Wangari Maathai, *The Green Belt Movement: Sharing the Approach and the Experience* (New York: Lantern Books, 2003), 38.

14. See especially Vandana Shiva, *Earth Democracy: Justice, Sustainability, and Peace* (Boston: South End Press, 2005), and *Soil Not Oil: Environmental Justice in an Age of Climate Crisis* (Boston: South End Press, 2008).

15. Wes Jackson, "The Agrarian Mind: Mere Nostalgia or a Practical Necessity?," in *The Essential Agrarian Reader,* edited by Norman Wirzba (Washington, DC: Shoemaker and Hoard, 2003), 141.

16. Ngugi wa Thiong'o, *Petals of Blood* (New York: Penguin, 1977).

17. For a related and insightful discussion of what she calls "articulated categories," see Anne McClintock, *Imperial Leather: Race, Gender and Sexuality in the Colonial Contest* (New York: Routledge, 1995), 4–6.

18. Ato Quayson, *Calibrations* (Minneapolis: University of Minnesota Press, 2003), 73.

19. Meear Selva, "Wangari Maathai: Queen of the Greens," *Guardian,* October 9, 2004, 9.

20. William Finnegan, "The Economics of Empire: The Washington Consensus," *Harper's,* May 2003, 48.

21. Bertolt Brecht, "An die Nachgeborenen" ["To Posterity"], in *Selected Poems,* translated by H. R. Hays (New York: Grove, 1959), 173.

22. The time frame here is crucial. With the help of international donors, Maathai put in place a system whereby each woman was paid a modest amount not for planting a tree but instead for keeping it alive for six months. If it was still growing at that point, she would be remunerated. Thus, the focus of the group's activities was not the single act of planting but rather maintaining growth over time. The literature on desertification is complex and conflicted, largely around questions of the scale and source of the problem as well as the quality of the research. Given the fraught debates over the implications

of desertification, I have avoided the term, preferring simply to reference the slow violence of soil erosion and deforestation. For two useful accounts of the spread of positions on this issue, see Jeremy Swift, "Desertification: Narratives, Winners and Losers," in *The Lie of the Land. Challenging Received Wisdom on the African Environment,* edited by Melissa Leach and Robin Mearns, 71–94 (Oxford, UK: James Currey, 1996), and William M. Adams, "When Nature Won't Stay Still: Conservation, Equilibrium and Control," in *Decolonizing Nature: Strategies for Conservation in a Post-Colonial Era,* edited by William M. Adams and Martin Mulligan, 220–46 (London: Earthscan, 2003).

23. Michela Wong, *It's Our Turn to Eat: The Story of a Kenyan Whistle-Blower* (New York: Harper 2009), 37.

24. Michael Pollan, *Second Nature: A Gardener's Education* (New York: Dell, 1991), 194.

25. Ramachandra Guha, *Environmentalism: A Global History* (New York: Longman, 2000), 115–24.

26. A major precursor to the conflict over Karura had occurred in 1989. The regime had been steadily appropriated and privatizing parts of Nairobi's Uhuru Park, which Maathai has likened to New York's Central Park and London's Hyde Park as a vital green space, a space for leisure and for political gatherings. When Maathai learned that the ruling party was to erect a sixty-story skyscraper for new party headquarters and a media center in Uhuru Park, battle was joined. Green Belt activists spearheaded a successful movement to turn back the regime's efforts to privatize public land under the deceptively spectacular iconography of national development. The regime would not forgive Maathai for humiliating them in this manner.

27. Amitabh Pal, "Maathai Interview," *Progressive,* May 2005, 5.

28. I've adapted the phrase from Al Gedicks, *Resource Rebels: Native Challenges to Mining and Oil Corporations* (Cambridge, MA: South End Press, 2001).

29. For the history of the forest fighters, see Caroline Elkins, *Imperial Reckoning: The Untold Story of Britain's Gulag in Kenya* (New York: Holt, 2005); Wunyabari O. Maloba, *Mau Mau and Kenya: An Analysis of a Peasant Revolt* (Bloomington: Indiana University Press, 1998); and especially David Anderson, *Histories of the Hanged: The Dirty War in Kenya and the End of Empire* (New York: Norton, 2005), 230–88.

30. For the most comprehensive discussion of this literature, see David Maughan-Brown, *Land, Freedom and Fiction: History and Ideology in Kenya* (London: Zed Books, 1985).

31. The Mau-Mau Revolt was far from being an undivided revolt: numerous fault lines opened up at times, not least between educated nationalist leaders and the predominantly peasant forest fighters.

32. Byron Caminero-Santangelo, "Different Shades of Green: Ecocriticism and African Literature," in *African Literature: An Anthology of Criticism and Theory*, edited by Tejumola Olaniyan and Ato Quayson (Oxford, UK: Blackwell, 2007), 702.

33. In many Kenyan novels about the Mau Mau period, the forest fighters are depicted with a cloying if understandable romanticism. Regarding the complex and varied legacies of colonial cultures of nature, it is worth noting that Maathai admires the Men of the Trees, an organization founded in Kenya in the 1920s that brought together British and Kikuyu leaders to promote tree planting (Maathai, *Unbowed*, 131).

34. Although the initial resistance came from the Green Belt Movement, the resistance spread to the streets of Nairobi, where it was taken up by a broad swath of the population, particularly students, both female and male.

35. See William Beinart and Peter Coates, *Environmental History: The Taming of Nature in the U.S.A. and South Africa* (London: Routledge, 1995); Jane Carruthers, *The Kruger National Park: A Social and Political History* (Pietermaritzburg, South Africa: Allen, 1995); and A. Fiona D. Mackenzie, "Contested Ground: Colonial Narratives and the Kenyan Environment, 1920–1945," *Journal of Southern African Studies* 26, no. 4 (2000): 697–718.

36. James C. Scott, *Seeing Like a State: How Certain Schemes to Improve the Human Condition Have Failed* (New Haven, CT: Yale University Press, 1998), 264.

37. Mackenzie, "Contested Ground," 27. Mackenzie, like Beinart, stresses that there were among colonial officialdom some dissident voices who recognized the value and applicability of local agricultural and environmental knowledge.

38. Stuart Jeffries, "Kenya's Tree Woman," *Mail and Guardian*, February 28, 2007, 7.

39. For an ambitious experiment in adaptively blending a singular self, a social movement, and a polemical agenda, see W. E. B. Du Bois, *Dusk of Dawn: An Essay toward an Autobiography of a Race Concept* (1940; reprint, New York: Transaction Publishers, 1983).

40. An important distinction should be made between the routes that Carson and Maathai took to their writing and their activism. Carson was a lifelong writer who remade herself as an activist late in life after she traded her lyrical voice (which she'd honed as a celebrant of marine life) for the voice of elegy and apocalypse in *Silent Spring*. Maathai's trajectory was in the opposite direction. An activist all her adult life, she became a writer of testimony only in her later years.

41. Linda Lear, *Rachel Carson: Witness for Nature* (New York: Holt, 1997), 254.

42. Quoted in ibid., 429.

43. "Pesticides: The Price of Progress." *Time,* September 28, 1962, 45, quoted in Lear, *Rachel Carson,* 461. Additionally, a review by Carl Hodge was titled "*Silent Spring* Makes Protest Too Hysterical," *Arizona Star,* October 14, 1962, 7.

44. Quoted in Lear, *Rachel Carson,* 417.

45. Quoted in ibid., 409.

46. Quoted in "The Silent Spring of Rachel Carson," Transcript, CBS Reports, April 3, 1963.

47. DAD, "Controversial Book by Rachel Carson Lives Up to Advance Warnings," *Aerosol Age,* October 1962, 81. I'm grateful to Lindsay Woodbridge for first drawing my attention to this review in her fine unpublished senior thesis "The Fallout of *Silent Spring*" (University of Wisconsin–Madison, 2007).

48. This misogyny, together with the regime's authoritarian intolerance of dissent, had profound professional and financial repercussions for Maathai. In 1982 after teaching at the University of Nairobi for sixteen years, she decided to run for parliament. To do so, she was told that she had to resign from her job at the university. She was then promptly informed by the electoral committee that she was disqualified (on a trumped-up technicality) from running for parliament. Twelve hours after resigning as chair of the university's Department of Veterinary Anatomy, Maathai asked for her job back. Under pressure from the regime, the university refused to reemploy her, denying her moreover all pension and health benefits. Maathai, a forty-one-year-old single mother with no safety net, was thrown out onto the streets. In 2005 shortly after Maathai was awarded the Nobel Prize, the very university that had treated her so appallingly tried to cash in on her international fame by awarding her an honorary doctorate in science.

49. Selva, "Wangari Maathai," 8.

50. Quoted in Jim Motavelli, "Movement Built on Power of Trees," *E: The Environmental Magazine,* July–August 2002, 11.

51. Ibid., 111. For a more elaborate account of the burden of traditionalism placed on women in the context of a Janus-faced modernity, see McClintock, *Imperial Leather,* 294–300.

52. Maathai, *Unbowed,* 115, 196. There are echoes between the nativist arguments mounted against Maathai by President Moi and the arguments of her ex-husband, Mwangi, who testified in court that he was divorcing her because she was ungovernable: "too educated, too strong, too successful, too stubborn, and too hard to control" (quoted in *Unbowed,* 146).

53. Kwame Anthony Appiah, *In My Father's House* (New York: Oxford University Press, 1992), 55.

54. Quoted in Patrick E. Tyler, "In Wartime, Critics Question Peace Prize for Environmentalism," *New York Times,* October 10, 2004, A5. Morten Hoeglund,

a member of Norway's Progress Party, concurred with Hagen, arguing that "the committee should have focused on more important matters, such as weapons of mass destruction," (quoted in Selva, "Wangari Maathai," 9).

55. Alister Doyle, "Kenyan Green Activist Wins Nobel Peace Prize," http:// www.ogiek.org/indepth/break-wang-nobel-pr.htm. See, for example, Maathai's insistence that through a focus on reforestation and environmental resource management, "we might preempt many conflicts over the access and control of resources" (*Unbowed*, xvi).

56. In Kenya, which boasts some forty ethnicities, the sources of ethnic tension are complex but have often been especially explosive along the fault lines between pastoralists and farmers where resources are overstressed. Divisive politicians have manipulated these tensions to their advantage during, for instance, the violence that beset the Rift Valley, Nyanza, and western provinces in the early 1990s and, more broadly, during the aftermath of the disputed national elections of 2007. The slow violence of resource depletion, a mistrust of government, and political leaders who play the ethnic card can easily kindle an atmosphere of terror that fuels social unrest.

57. Quoted in Ofeibea Quist-Arcton, "Maathai: Change Kenya to Benefit People," www.greenbeltmovement.org (accessed January 1, 2003).

Bibliography

Adams, William M. "When Nature Won't Stay Still: Conservation, Equilibrium and Control." In *Decolonizing Nature: Strategies for Conservation in a Post-Colonial Era*, edited by William M. Adams and Martin Mulligan, 220–46. London: Earthscan, 2003.

Anderson, David. *Histories of the Hanged: The Dirty War in Kenya and the End of Empire*. New York: Norton, 2005.

———. "Kenya's Agony." Royal Africa Society On-line, January 10, 2008, http://www.royalafricansociety.org/index.php?option=com_content& task=view&id=443.

Appiah, Kwame Anthony. *In My Father's House*. New York: Oxford University Press, 1992.

Beinart, William, and Peter Coates. *Environmental History: The Taming of Nature in the U.S.A. and South Africa*. London: Routledge, 1995.

Branch, Daniel. "At the Polling Station." *London Review of Books*, January 24, 2008, 26–27.

Brown, Lester R. "The Price of Salvation." *Guardian*, April 25 2007, 1–4.

Caminero-Santangelo, Byron. "Different Shades of Green: Ecocriticism and African Literature." In *African Literature: An Anthology of Criticism and Theory,* edited by Tejumola Olaniyan and Ato Quayson, 698–706. Oxford, UK: Blackwell, 2007.

Carruthers, Jane. *The Kruger National Park: A Social and Political History.* Pietermaritzburg, South Africa: Allen, 1995.

Carson, Rachel. *Silent Spring.* 1962; reprint, Boston: Houghton Mifflin, 1992.

Crichton, Michael. *State of Fear.* New York: Avon, 2004.

DAD. "Controversial Book by Rachel Carson Lives Up to Advance Warnings." *Aerosol Age,* October 1962, 81–82, 168.

Elkins, Caroline. *Imperial Reckoning: The Untold Story of Britain's Gulag in Kenya.* New York: Holt, 2005.

Finnegan, William. "The Economics of Empire: The Washington Consensus." *Harper's,* May 2003, 41–54.

Friedman, Thomas. "The Age of Interruption." *New York Times,* July 5, 2006, A27.

Gedicks, Al. *Resource Rebels: Native Challenges to Mining and Oil Corporations.* Cambridge, MA: South End Press, 2001.

Guha, Ramachandra. *Environmentalism: A Global History.* New York: Longman, 2000.

Guha, Ramachandra, and Juan Martinez-Alier. *Varieties of Environmentalism: Essays North and South.* London: Earthscan, 1997.

Hertzberg, Hendrik. "The War in Iraq." *New Yorker,* March 27, 2003, 15.

Hodge, Carl. "*Silent Spring* Makes Protest Too Hysterical." *Arizona Star,* October 14, 1962, 14.

Jackson, Wes. "The Agrarian Mind: Mere Nostalgia or a Practical Necessity?" In *The Essential Agrarian Reader,* edited by Norman Wirzba, 140–53. Washington, DC: Shoemaker and Hoard, 2003.

Jeffries, Stuart. "Kenya's Tree Woman." *Mail and Guardian,* February 28, 2007, 7–8.

Lappe, Anna, and Frances Moore Lappe. "The Genius of Wangari Maathai." *International Herald Tribune,* October 15, 2004, 1–2.

Lear, Linda. *Rachel Carson: Witness for Nature.* New York: Holt, 1997.

Maathai, Wangari. *The Green Belt Movement: Sharing the Approach and the Experience.* New York: Lantern Books, 2003.

———. *Unbowed: A Memoir.* New York: Knopf, 2006.

Mackenzie, A. Fiona D. "Contested Ground: Colonial Narratives and the Kenyan Environment, 1920–1945." *Journal of Southern African Studies* 26, no. 4 (2000): 3–33.

Maloba, Wunyabari O. *Mau Mau and Kenya: An Analysis of a Peasant Revolt.* Bloomington: Indiana University Press, 1998.

Maughan-Brown, David. *Land, Freedom and Fiction: History and Ideology in Kenya.* London: Zed Books, 1985.

McClintock, Anne. *Imperial Leather: Race, Gender and Sexuality in the Colonial Contest.* New York: Routledge, 1995.

Motavelli, Jim. "Movement Built on Power of Trees." *E: The Environmental Magazine,* July–August 2002, 11–13.

Ngugi wa Thiong'o. *A Grain of Wheat.* London: Heinemann, 1967.

Nyambura-Mwaura, Helen. "Seeds of Class War Sprout in Kenya's Crisis." *Mail and Guardian,* February 13, 2008, 15–16.

Pal, Amitabh. "Maathai Interview." *Progressive,* May 2005, 3–5.

"Pesticides: The Price of Progress." *Time,* September 28, 1962, 45.

Pollan, Michael. *Second Nature: A Gardener's Education.* New York: Dell, 1991.

Quayson, Ato. *Calibrations.* Minneapolis: University of Minnesota Press, 2003.

Quist-Arcton, Ofeibea. "Maathai: Change Kenya to Benefit People." AllAfrica.com, January 1, 2003, http://www.greenbeltmovement.org/a.php?id=130.

Schecter, Arnold, et al. "Agent Orange and the Vietnamese: The Persistence of Elevated Dioxin Levels in Human Tissues." *American Journal of Public Health* 85, no. 4 (1995): 516–22.

Scott, James C. *Seeing Like a State: How Certain Schemes to Improve the Human Condition Have Failed.* New Haven, CT: Yale University Press, 1998.

Selva, Meear. "Wangari Maathai: Queen of the Greens." *Guardian,* October 9, 2004, 7–9.

Sevareid, Eric. "An Explosive Book." *Washington D.C. Star,* October 9, 1962, 3.

"The Silent Spring of Rachel Carson." Transcript, CBS Reports, April 3, 1963.

Serumaga, Kalundi. "Unsettled—Hitting without Touching." *Liberation Lit,* February 18, 2008. liblit.wordpress.com/2008/02/18/unsettled- by-kalundi-serumaga.

Swift, Jeremy. "Desertification: Narratives, Winners and Losers." In *The Lie of the Land: Challenging Received Wisdom on the African Environment,* edited by Melissa Leach and Robin Mearns, 71–94. Oxford, UK: James Currey, 1996.

Tyler, Patrick E. "In Wartime, Critics Question Peace Prize for Environmentalism." *New York Times,* October 10, 2004, A5.

———. "Peace Prize Goes to Environmentalism in Kenya." *New York Times,* October 9, 2004, A1, A5.

United States Department of Defense. "DoD News Briefing—Secretary Rumsfeld and Gen. Myers." March 21, 2003.

Vasagar, Jeevan. "Bulldozers Go in to Clear Kenya's Slum City." *Guardian,* April 20, 2004, 6.

"Vietnam in Retrospect." *New York Times,* March 23, 2003, A25.

Woodbridge, Lindsay. "The Fallout of *Silent Spring.*" Senior thesis, University of Wisconsin–Madison, 2007.

Contributors

Byron Caminero-Santangelo is associate professor of English at the University of Kansas. He is author of *African Fiction and Joseph Conrad: Reading Postcolonial Intertextuality* (SUNY Press, 2005). Some of his recent scholarship includes "Shifting the Center: A Tradition of Environmental Literary Discourse from Africa" *Environmental Criticism for the Twenty-First Century* (Routledge 2011); "In Place: Tourism, Cosmopolitan Bioregionalism, and Zakes Mda's *The Heart of Redness*" *Postcolonial Ecologies: Literatures of the Environment* (Oxford UP 2011); and "Different Shades of Green: Ecocriticism and African Literature" *African Literature: An Anthology of Criticism and Theory* (Blackwell 2007).

Jane Carruthers is professor of History at the University of South Africa and Chair of the Academic Advisory Board of the Rachel Carson Center. She has published extensively and her book, *The Kruger National Park: A Social and Political History,* is well known and widely used. She is a Fellow of the Royal Society of South Africa and a Fellow of Clare Hall, Cambridge. Her current research interests lie in the fields of the history of the biological sciences and national parks, colonial art, heritage and cartography. Jane has also been involved in research on land restitution claims and land reform and national park initiatives in South Africa and Australia.

Mara Goldman is assistant professor of Geography and a research associate in the Environment and Society Program at the Institute of Behavioral Science at the University of Colorado at Boulder. Mara's research has focused on the interface of human-environment relations and critical geographies of conservation and development, with a regional focus in East Africa. Her work addresses the politics of knowledge and participation as related to wildlife conservation interventions and rangeland management, changing pastoral livelihood and communication practices, and empowerment and governance issues within Maasai communities in Tanzania and Kenya.

Amanda Hammar is professor of African Studies in the Centre for African Studies, University of Copenhagen, Denmark. She co-edited (with Brian Raftopoulos and Stig Jensen) *Zimbabwe's Unfinished Business: Rethinking Land, State and Nation in the Context of Crisis* (Weaver 2003). She has authored and co-authored numerous journal articles in *Journal of Southern African Studies, Journal of Contemporary African Studies,* and *Journal of Agrarian Change*, while also co-editing two recent special issues of the first two of these journals around themes of (political economies of) displacement in southern Africa. She has also written a number of book chapters dealing with questions of land, sovereignty and displacement in southern Africa.

Jonathan Highfield is professor of English at Rhode Island School of Design. He is the author of "'Relations with Food': Agriculture, Colonialism, and Foodways in the Writing of Bessie Head" *Postcolonial Green: Environmental Politics and World Narratives* (2010); "Driving the Devil into the Ground: Settler Myth in Andre Brink's *Devil's Valley*" *Trauma, Resistance, Reconstruction in Post-1994 South African Writing* (2010); and "'A Breath Out of the Heart of the Country': The Landscapes of David Malouf" *Fact and Fiction: Readings in Australian Literature* (2008). He is the co-editor (with Kwadwo Opoku-Agyemang and Dora Edu-Buandoh) of *The State of the Art(s): African Studies and American Studies in Comparative Perspective* (2006). He is currently working on a project on foodways and literature in Africa.

David McDermott Hughes is associate professor in the Department of Anthropology, Rutgers University. He is the author of *Whiteness in Zimbabwe: Race, Landscape and the Problem of Belonging* (Palgrave Macmillan 2010) and *From Enslavement to Environmentalism: Politics on a Southern African Frontier* (University of Washington Press 2006). His works straddle the boundary between anthropology and geography and are focused on Southern Africa, particularly Zimbabwe.

Garth Myers is Paul E. Raether Distinguished Professor of Urban and International Studies in the Center for Urban and Global Studies and Department of International Studies at Trinity College, Hartford, CT. He is the author of *African Cities: Alternative Visions of Urban Theory and Practice* (Zed Books 2011), *Verandahs of Power: Colonialism and Space in Urban Africa* (Syracuse 2003), and *Disposable Cities: Garbage, Governance and Sustainable Development in Urban Africa* (Ashgate 2005), and co-editor with Martin J. Murray of *Cities in Contemporary Africa* (Palgrave 2006).

Roderick Neumann is professor in the Department of Global and Sociocultural Studies at Florida International University. His scholarship is organized around two lines of inquiry: one, the co-constitution of nature, society, and landscape; and two, the political economy of the environment. Underlying these lines of inquiry are normative concerns for social justice and biodiversity and habitat protection. He is the author of *Making Political Ecology* (Hodder Arnold 2005) and *Imposing Wilderness* (University of California Press 1998), as well as more than two dozen research articles and book chapters.

Rob Nixon is Rachel Carson Professor of English at the University of Wisconsin-Madison. He is the author of four books: *Slow Violence and the Environmentalism of the Poor* (Harvard University Press 2011); *London Calling. V. S. Naipaul, Postcolonial Mandarin* (Oxford); *Homelands, Harlem and Hollywood. South African Culture and the World Beyond* (Routledge); and *Dreambirds: the Natural History of a Fantasy* (Picador), selected by the *New York Times Book Review* as a Notable Book of the Year and by *Esquire* as one of the ten best books of the year. Professor Nixon is a frequent contributor to the *New York Times*; his writing has also appeared in publications such as *The New Yorker, Atlantic Monthly, London Review of Books, Times Literary Supplement, Village Voice, The Nation, The Guardian,* and *Outside.*

Anthony Vital is professor of English at Transylvania University, a liberal arts college in Kentucky. In the late 1990s he began presenting conference papers that read South African literature in relation to South Africa's developing culture of environmentalism, one transformed by the transition to democracy. From these papers, which keep attention focused on the country's environmental justice movement, he has been publishing articles exploring a literary interpretation that foregrounds issues relating ecology, environmental justice and the postcolonial condition.

Laura Wright is associate professor and director of Graduate Studies in English at Western Carolina University, where she specializes in postcolonial literature and theory. She has written two books, *Writing "Out of All the Camps": J. M. Coetzee's Narratives of Displacement* (2006) and *Wilderness into Civilized Shapes: Reading the Postcolonial Environment* (2010). She is co-author (with Elizabeth Heffelfinger) of a third, *Visual Difference: Postcolonial Studies and Intercultural Cinema* (2011) and lead editor (with Jane Poyner and Elleke Boehmer) of the approved MLA volume *Approaches to Teaching Coetzee's Disgrace and Other Works.*

Index

CPSIA information can be obtained at www.ICGtesting.com
Printed in the USA
LVOW07s2232090115

422154LV00001B/4/P